SMART HOMES AND COMMUNITIES

智能社区

SMART HOMES AND COMMUNITIES

智能社区

[加] 阿维·弗里德曼 编著
齐梦涵 译

广西师范大学出版社
· 桂林 ·
images Publishing

图书在版编目(CIP)数据

智能社区/(加)阿维·弗里德曼编著;齐梦涵译.—桂林:广西师范大学出版社,2018.5
ISBN 978-7-5598-0724-3

Ⅰ.①智… Ⅱ.①阿… ②齐… Ⅲ.①智能化建筑-建筑设计
Ⅳ.①TU243

中国版本图书馆 CIP 数据核字(2018)第 053108 号

出 品 人:刘广汉
责任编辑:肖 莉
助理编辑:齐梦涵
版式设计:吴 迪

广西师范大学出版社出版发行
(广西桂林市五里店路9号 邮政编码:541004
网址:http://www.bbtpress.com)
出版人:张艺兵
全国新华书店经销
销售热线:021-65200318 021-31260822-898
恒美印务(广州)有限公司印刷
(广州市南沙区环市大道南路334号 邮政编码:511458)
开本:787mm×1 092mm 1/12
印张:21 字数:30千字
2018年5月第1版 2018年5月第1次印刷
定价:256.00元

目录

6 序

10 第一章：自然思维
14 奇卡诺社区
18 穿越草原社区
22 生态现代公寓
26 石楠木教堂社区

30 第二章：移动思维
34 8号公寓
40 奥伦科车站社区
44 假日近邻社区
48 昆次伏广场社区

52 第三章：基础设施与住宅的混合思维
56 前景新城
60 马斯达尔城市开发区
66 大学城
70 港池7号

74 第四章：绿色开放空间
78 海勒街公园和住宅
82 温嫩登乐土社区
86 布里克近邻社区
90 阿尔博莱拉生活社区
94 协和社区

98 第五章：密集型住宅小区
102 克里斯蒂・沃克社区
106 丐军庭院社区
110 帕克兰公寓
114 欧塞奇庭院社区

118 第六章：新旧混合
122 艺术风景韦奇伍德谷仓社区
126 24号巷
130 拉格公寓
134 伊顿天然社区
138 卡恩・里巴斯工厂改造项目

142 第七章：可食用景观
146 青草巷公寓
150 沃德维尔庭院公寓
154 比弗兵营社区住房
158 占地社区
162 宽街18号社区

166 第八章：区域供热
170 朱比利码头社区
174 弗莱堡太阳能拓居地
178 中央车站铁马社区
182 奥斯特朗社区
186 古什迪恩太阳能建筑

190 第九章：创新住宅概念
194 多尼布鲁克小区
198 郊区复兴生活区
202 格林威治千年村第二期
206 格林别墅区

210 第十章：低碳住宅
214 哥伦比亚车站微型社区
220 奥斯丁SOL社区（1.0版）
224 甜水光谱社区
228 同胞绿色社区
234 z之家社区

238 致谢
239 章节参考书目
242 项目参考书目
247 照片版权信息
248 建筑师信息
251 索引

序

用“智能”这个词来形容住宅和社区是于 20 世纪 90 年代创造出来的，人们用它来描述某些场所的功能与信息技术有关或受到信息技术的影响而产生。在数字革命早期，设计师与规划师们必须在设计居住空间与社区空间的同时，也要考虑到新的经济现实与创新数字家电。

信息技术虽然在我们的活动中占据主导位置，但是随着几十年来社会关注点的不断变化发展，其所处的优先地位已发生了转变，人们不得不去关注其他的新生现象。当前的住宅和社区规划设计模式正面临着原理和模式的双重挑战，这一点已经毋庸置疑。过去的方法已经不能满足现代的需求，所以我们更需要创新思维。环境、经济和社会等方面的基本原则都发生了变化，这也推动了人们对一种新的前景的需求。

不可再生自然资源的枯竭、温室气体排放量的增加和气候变化等环境挑战，迫使设计师不得不重新思考设计的概念与方法，以期在社区与自然之间获得更好的兼容。建筑师和建筑商开始在自己的思维过程和住宅设计实践中整合融入一些当代策略，这其中包括最大限度地减少发展中产生的碳排放量、区域供热、零能耗住宅和保护当地自然资产的总体规划概念。

材料、劳动力、土地和基础设施的成本持续上涨，这给设计师们带来了经济上不小的挑战，支付能力成了重中之重。对更少投入的需求产生了集约型社区、可更改与可扩张型住宅，以及小型精良住房的概念。另外，对减少公共设施费用的需求促成了更好的建筑实践，这对环境与居住者都有好处。

社会挑战也引起设计师、建筑商和房主的注意。随着婴儿潮一代人的退休，老年人的居住问题成为首要问题。步行社区、原居安老和多代同堂住宅成为设计师们需要考虑的概念。此外，对于那些希望在家工作的人来说，居家办公环境也已经成为经济现实的一部分——这种工作方式通过数字技术的进步已经成了可能。

对以创新思维思考居民社区的需求使我萌生了创作本书的想法。本书的目的在于为读者提供当代社区与住宅设计概念的信息，并通过出色的国际案例来对这些概念进行说明。因此，智能建筑在本书中被视为旨在解决上述新问题的场所，能够有效应对新出现的这些挑战。而本书中经常被提及的另一个词语则是可持续性，这也是本书内容的一个基本原则。

“可持续发展”这个术语的扩散及其产生的条件可以追溯到几十年前。1972 年，联合国人类环境会议在瑞典斯德哥尔摩举行，主要探讨了人类的发展正在考验“地球的承载能力”的问题。这次会议是首个对正在进行的环境破坏与人类未来之间的关系进行讨论的国际会议。人们意识到，一些国家的人口增长和过度消费导致了土地退化、森林面积减少、空气污染和水资源短缺等一系列问题。

几年后，这种反思促进了世界环境与发展委员会（WCED）的成立，这是一项全球计划。在 1987 年的“共同未来”这一报告中，作者将可持续发展定义为“既能满足我们现今的需求，又不损害子孙后代，还能满足他们的需求的发展模式”。这一定义确立了新时期人类发展的方法，即采取任何行动时都应考虑其对未来的影响。委员会还制定了一套发展模式，其主要标准是对社会公平的要求，在国家内部及国家之间公平分配资源的要求，以及解决发展压力和环境所造成的冲突的需求。随着时间的推移，这些基础问题及其相互之间的关系已经成为衡量一切发展活动是否成功的标准。

社会、经济和环境是可持续发展的三大支柱。人们逐渐发现，如果想要成功地实施可持续发展计划，文化与治理也是必不可少的。对文化这一方面的担忧反映出了居民的社会需求以及他们的价值取向。社会需求和公平是一个广泛而全面的概念，可以从多种角度进行解释。例如，如果我们的目标是建立一个可持续的医疗保健体系，那么我们必须确保源源不断的充足资金。鼓励健身也可以促进公共健康。有证据显示积极的生活方式能减少人们患上心血管疾病与糖尿病等疾病的可能性。因此，从城市的最佳利益出发，社区中应该包含自行车道和人行道，住宅和非住宅的功能也应该被整合在一起。

弘扬乡土文化、保护地方传统、保护文物建筑，也会以或直接或间接的方式为社会做出贡献。值得保护的旧建筑物是人们看得见的对人类历史的提醒。尊重历史的人或许更愿意为提升未来建筑物的质量贡献力量。对旧建筑的保护和改造也能避免拆迁，有助于减少对自然资源的消耗。

促进经济可持续发展是城市规划的另一个目标。其目的是为了避免把我们现在所做的错误决定产生的不良后果强加给我们的后代子孙。例如，建设不必要的宽阔马路，而不是普通的狭窄街道，这可能会对经济产生长期的影响，因为街道需要定期重新铺设，而在气候寒冷的地区，更宽的马路也会堆积更多的积雪，需要更高的人力成本去除雪。在私人开发项目中，较宽的道路费用会提高每栋房屋的价格，可能会迫使买家借更多的钱，他们将不得不长期偿还这些债务，从而使自身的财务可持续性面临风险。

环境可持续发展关注的是建设和发展创建的生态属性，包括开发区中的道路、开放空间和住宅。规划一个开发区时，进行可持续发展的评估是必不可少的。它不仅会影响建筑师对材料的选择，还会影响到所使用的材料的长期性能以及这些材料在使用结束之后的可回收性。沥青路面会导致雨水汇聚成水流直接流向排水孔，在道路两旁建造生态水道可以促进雨水植物的生长，进而减少雨水径流。

治理是可持续发展的另一个重要的方面。可持续发展的战略和概念尽管具有创新性，但是仍然需要市政领导制定适当的政策并向公民解释其长远的发展，否则便不能顺利实施。一个有效的政治制度也将吸引新的参与者和年轻人参加到这种公共服务中来，产生一种思想上和行动上的连续性。

可持续发展的五大支柱都是独立的。然而，仔细审视以可持续性发展原则为基础规划建设的社区的内部运作，我们不难看出这几个方面之间的通力合作对建筑环境有至关重要的影响。而这种通力合作正是本书的重点。这五个重叠的方面相互作用，形成了一系列概念及解决方法，而认识到这些方面的相互关系则是非常重要的。

对城市环境和住宅的设计是其子构件的聚集和分层。一个运转良好的城市是其交通、经济、治理、环境及其他各个问题综合考虑并解决的结果。通过考察这些因素以及它们之间如何相互作用，才能使人们将城市作为整体来欣赏。

本书分别审视了上述每一个元素，以期了解它的重要性和工作原理，同时牢记它是如何与更广阔的视角相关联的。

第一章着眼于建筑与自然环境的关系。建造社区的位置及其朝向可能会提高或降低其可持续性。必须在设计过程初期就认识到选址的意义，这样才能确保自然和住宅的成功整合。该章讨论的正是关键性的规划考虑因素，包括气候、被动式太阳能获取、风向、动植物保护、池塘和地形。

第二章主要着眼于流通性。设计出布局合理的街道、自行车和人行道组成的流通网络是规划可持续社区的中心。该章讨论了怎样为新开发的社区设计道路、停车场和街道模式最为合理，并解释了它们对交通的影响和由此产生的街道外观吸引力。

公共设施和住宅的混合是第三章的主题。把家庭生活和非家庭生活的活动隔离开来进行设计是主流的设计趋势，这种做法加强了人们对私家车的依赖，并且需要建立一个广泛的道路网络。事实上，商业及其他社区功能，例如图书馆和诊所，可以被设置在步行能够到达的社区中心位置，或者被设置在新开发社区的边缘，以供临近社区的居民造访，这也有利于在经济方面支持这些公共设施。

第四章主要讨论的话题是如何在考虑到居民生活方式的同时规划绿色休憩用地。社区居民年龄层和行动能力各不相同，因此设计出一个不仅适合儿童，也适合喜爱活动的成年人及老人的场所至关重要。此外，一个良好的开放空间可以为居民提供多种选择，以适应人们不同的情绪和想要参加的活动。一个开放空间应该包括能够为出售旧货而举办的集市提供坚硬路面的场地，也包括为休闲游戏而建的路面柔软的场所。该章探讨了支持积极的生活方式的创新开放空间设计，并用最近的社区设计案例对其进行了说明。

减少社区对生态的影响是第五章的主题。为了不断适应变化的城市环境，我们把碳排放总量较小的密集型社区模式纳入了考量。本章回顾了高密度生活的演变历史，阐述了当代的创新战略、规划宜居的紧凑型社区的指导原则，以及我们在现实世界中将其实施的方法。

第六章将社区本身的历史传承与其可持续发展联系在一起，我们从中能够发现社会、经济和环境等多方面的问题。该章探讨了文物保护的进化发展，以在现有住宅用地上加建新房屋的大型住宅社区为例，说明了填充式房屋的设计原则。

为了应对食品成本的上涨，满足人们对获得营养和减少交通对环境影响的需求，第七章讨论了可食用景观社区。这些社区为居住者提供了可以亲自在自己的土地上或指定的公共花园里种植可食用植物的机会。该章还介绍了创新型都市农业实践方法和实践案例。

当整个社区的供热都来自一个中央供热源时，这样的供热方式被称为区域供热。供热的能量来源可以是太阳能、风能，也可以是地热或天然气等能源。这种系统的优点是可以为每个家庭节约供暖成本，提高能源的利用效率，降低碳排放量。该章讨论了区域供热的原则，说明了其在多个社区中的开创性实施。

一些人认为可持续发展就是给住宅加装“绿色的”科技设备。但是一所住宅是否是可持续性的还取决于它的形式和功能，即它的预期用途是否都被有效地利用了。具备创新的居住理念能够更好地促进社会、经济和文化等方面的可持续发展，而第九章就展示了具备这样的居住理念的社区，其中包括可动住宅、多代聚居住宅、插件单位住房和即插即用住房等。

能源成本上升，自然资源日益减少以及对居民健康的担忧，促使人们重新思考住宅的设计与建造。因此，积极应对环境挑战的住房在房屋市场上迅速普及。第十章首先从早期文明的居住方式中寻求解决办法，然后对这种设计原则和"绿色"住宅的案例进行概括，再将其与此类社区结合起来进行阐释。

当设计的综合方法包括上述讨论的多个主题时，居住环境的可持续发展就有可能实现。不过，只思考及应用其中的某一个方面也会有所作为，就这一点而言，本书旨在为实现这一目标做出贡献。

第一章：自然思维

CONSIDERING NATURE

When planning a smart community, one of the goals needs to be protecting the site's biodiversity while at the same time creating a livable place. Among the areas under consideration will be sun path, wind direction, and preservation of topography, flora and fauna. These considerations contribute to improving the ecological, economic, and social attributes of a neighborhood. This chapter sets out the design principles of developments where the built and natural environments are well integrated.

The natural features of a site play a major role in its future vitality, the wellbeing of the residents, and a community's potential to become sustainable. For example, trees stabilize ground cover, prevent soil erosion, and absorb carbon dioxide. Research shows that fifteen trees are required to convert the carbon emission from one year of car use and about 40 trees are needed for a single home (Construction 2007).

The orientation of a community will affect its energy consumption. The appropriate orientation will enhance both its passive solar gain and increase the temperature from solar radiation (Egan 2013). Proper exposure to the sun also increases the entry of natural light. The daily and yearly positions of the sun should be evaluated early in the planning process by using sundials and path diagrams or computer simulation programs (Piedmont-Palladino and Mennel 2009). In the northern hemisphere the homes should be oriented towards the south to take advantage of the sun as much as possible. Rectangular houses are preferable for maximizing sun exposure. A rectangular house's ridgeline should be oriented on an east-west axis for maximum solar potential of the southern side (Piedmont-Palladino and Mennel 2009).

Based on the sun path, the majority of the windows should be located on the south side to absorb heat during the winter. During the summer, buildings will need to have shading devices to remain cool since the sun follows a higher path at that time of year. By having windows facing south and using a shading system during the summer, a house can reduce its heating and cooling costs by as much as 85 percent (Lea 2012).

Southern-oriented windows should be placed within five degrees of true south. As the degree increases, the overall solar efficiency of the house decreases. Note that these directions are given in reference to the sun and not the magnetic north. Existing vegetation such as trees can be used as shading devices during the warmer months. The trees will shade the house yet allow some light to penetrate through. During the winter period, when the leaves have fallen, the sun's rays will enter the home to increase the ambient temperature—saving energy and reducing heating costs. The age, species, growth rate, and mature canopy of the trees should be studied before planning on where to orient a home in relation to the sun (Gromicko 2013).

Another environmental factor when considering nature is the prevailing winds. Prevailing winds are winds that blow predominantly from a single, general direction over a particular point (Gromicko 2013). A region's prevailing winds are usually affected by the global pattern of movements in the Earth's surface. Wind roses are used to determine the direction of the winds. This is a graphic tool used by meteorologists to measure wind speeds and directions at a specific location. Wind roses present their data in a polar coordinate grid and plot different frequencies of winds blowing from different directions (Gilmer 2010). This tool helps in planning strategies to prevent soil erosion and destruction of agricultural lands. The information can also be used to prevent snow from piling up against windows and doors by calculating the wind patterns during winter. In the context of sustainable communities, the data collected from the wind roses is utilized to design a building that will take advantage of summer breezes for passive cooling, and protect against winds that can further chill the interior spaces during the winter season (Gromicko 2013).

Winds are a great source of natural ventilation in summer months. To allow the lowering of the ambient temperature, a community's streets need to be planned according to the direction of the wind channel, which differs in every region. In addition, favorable positioning of the vegetation will permit cross-ventilation so that mechanical cooling will not be necessary.

During the colder months, trees can act as a windbreak from the winter breezes. Natural, sustainable windbreaks are used as a protection against the wind, a controlling agent against the drifting of snow, and as an energy saver (Current 2011). Commonly in the northern hemisphere, cold winds come from northerly and westerly directions during winter. Consequently, coniferous trees are appropriate shields from the cold prevailing winds as their foliage stays all year long. The effectiveness of a windbreak will depend on the species' foliage density and on the spacing of the trees. For example, a single row of conifers will shield 40 to 60 percent of the cold prevailing winds, while multiple rows will block 60 to 80 percent of the cold winds (Gilmer 2010). Such trees should be planted along the northern and western side of a development to protect the houses from cold winds, therefore preventing the reduction of the indoor temperature. This organization will not interfere with the advantage of passive solar gain and summer breezes as the southern and eastern sides are not affected by winter breezes and do not need windbreaks. The trees should be planted approximately ten times the tree-to-house height ratio. Different tree species on diverse sides of the houses can help control the climate and save energy. As a result, a community can be both comfortable to live in and remain forested.

Ensuring that the site's groundcover will not be altered is also important. The roots of trees and shrubs that reach deep into the ground can easily be affected by soil compaction or water loss. During construction, the integrity of the site can be maintained by constructing retaining walls and terraces to reduce the chances of soil erosion. Erosions can disturb the site's natural drainage system and pollute nearby water sources.

Maintaining the site's natural cover is also important. Prior to construction, the condition of the trees should be evaluated to determine which should be removed or kept. Trees should be removed when they are dying or causing a major obstruction and mature trees should be protected. Mature trees are those that have reached 75 percent of their full canopy growth, determined by their girth measurement. For example, a maple tree has reached maturity when its circumference is 31 to 53 inches (78.74 to 134.62 centimeters) in diameter (Smallidge n.d.). During construction, the remaining trees need to be wrapped and guarded against damage as they are extremely important to the prosperity and wellbeing of the community.

For additional preservation of the flora, each road or path can be woven into the local surroundings and away from existing vegetation. In this way, ponds, rocks, and trees can be retained. There are potential cost savings to be gained by thinking ahead and planning for the fauna and biodiversity. Having green spaces and areas for wildlife will also add value to certain communities (Construction 2007). Biodiversity is also a useful means for engaging these communities and the industry, helping to strike a balance among the social, economic and environmental needs of sustainable developments.

The natural topography of the site should be retained so that there is minimal disturbance of the natural drainage. This is essential because the topography may not be able to be restored once it has been modified and this can change or damage the local ecosystem for good. The site's developments, therefore, need meticulous planning and topographical surveys to take advantage of its natural contours, minimize the amount of grading, and reduce the destruction of trees and natural drainage ways. A topographical survey would gather information about the natural and manufactured features of a site. The data consists of the position, level, and description of each feature.

Homes and roads should be sited according to natural slopes. The constructed homes should be above the street grade for appropriate drainage. A sloped construction site also promotes simple, natural, and effective drainage to reduce construction costs. Steep slopes should be avoided to keep down construction costs. The topography of gardens should ideally reflect the original relief to keep the impact on drainage patterns to a minimum, but bunds can sometimes be created to enhance visual or natural acoustic privacy.

In recent decades, it has become clear that human activity has damaged the natural integrity of ecosystems threatening the security of the societies that depend on these systems (Roseland 2000). A stronger emphasis should be made on maintaining such environmental assets for future generations. Consider passive solar gain, wind direction, preservation of flora and topography, these are major factors for the sustainability of a community. Recognize the implications of siting decisions early in the design process and ensure a successful integration of natural and the build environments as this will lead to the building of sustainable communities.

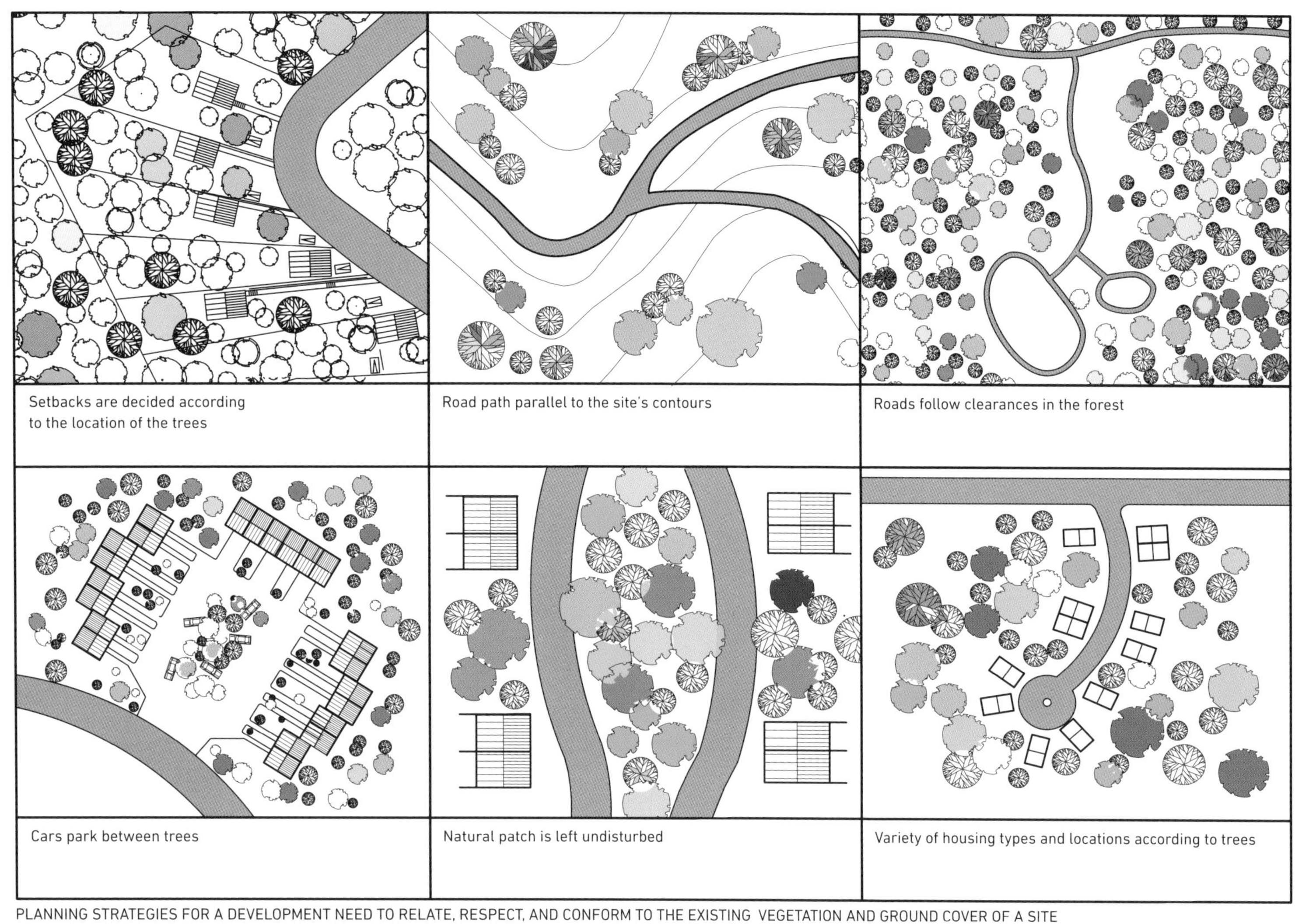

PLANNING STRATEGIES FOR A DEVELOPMENT NEED TO RELATE, RESPECT, AND CONFORM TO THE EXISTING VEGETATION AND GROUND COVER OF A SITE

奇卡诺社区
Civano

Tucson, Arizona, United States

DPZ CoDESIGN; Moule & Polyzoides, Architects & Urbanists; Wayne Moody

Civano is located in the southeast area of Tucson, Arizona and was defined by its planners as a remedy for the five consequences of urban sprawl. These five aspects include loss of community, loss of open space, traffic congestion, air pollution, and poor use of resources. In other words, Civano gave inhabitants the opportunity to choose how they wanted to live. Its design uniquely combines 600 housing units with communal spaces such as shopping, workplace, civic facilities, gathering places, commercial spaces, and natural open spaces making them part of the day-to-day life of its residents. In addition, sustainable energy techniques integrated in the design include passive and active solar principles, the use of eco-friendly building materials, and the implementation of water conservation systems.

With an area of 818 acres (331 hectares) of undeveloped desert in southeast Tucson, Civano adopted three rules that helped in conducting its land use accordingly, as well as its physical, social, and economic development. These three guidelines included: creating a sense of place that fosters communal values while connecting to the natural environment and with one another; minimizing harm to nature thanks to innovative design and using sustainable construction methods and materials; and implementing new technologies to augment quality of life. This unique land use is inspired by a brand new vision of community that harmonizes the balance of human needs and natural resources.

Civano creates the opportunity to regenerate nature by working with a sustainable ecological system. For example, due to the high percentage of Tucson's rainfall that is engineered to drain out of the county or is lost through evaporation, the Permaculture Drylands Institute wanted ways in which to retain Tucson's water in order to nourish vegetation and replenish wetlands. Civano allows this goal to be attained by preserving 35 percent of the neighborhood for natural and enhanced green open spaces reserved for communal orchards, linear parks, pedestrian pathways, bike trails, environmentally-friendly recreational facilities, and preserved desert wild lands.

The community also sets aside trees and plants by implementing a Civano Nursery Salvage Program to help preserve vegetation and educate residents on desert landscaping.

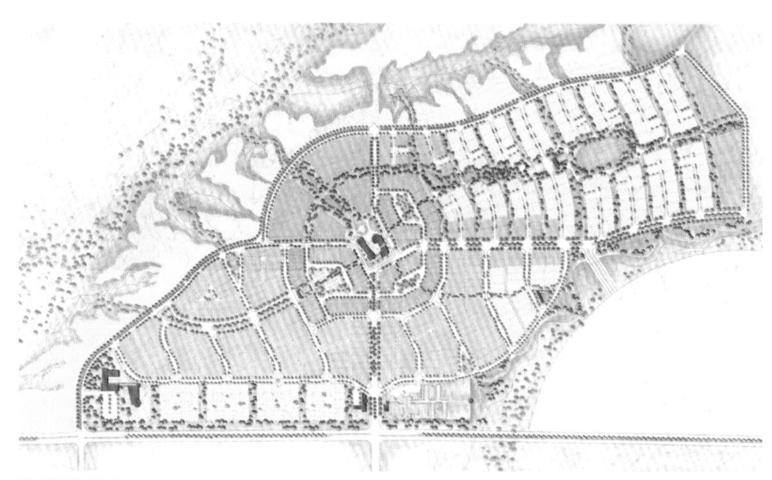

MASTER PLAN

A unique landscape design allows residents to connect with the natural environment where aspects of passive solar shading, permaculture, xeriscaping, and water harvesting techniques create a prosperous, sustainable community. Resource-efficient building techniques included using straw bales for insulation, employing passive solar (i.e. sun, wind, and shade), water harvesting and cooling, selecting recycled materials, active solar energy (i.e. solar photovoltaic panels and hot water systems located on roofs), cool tower, taking into account the thermal mass of the walls, and inserting highly efficient windows.

COMPOUND BLOCK SITE PLAN

COURTYARD HOUSING SITE PLAN

穿越草原社区

Prairie Crossing

Grayslake, Illinois, United States

Margaret McCurry FAIA, Tigerman-McCurry Architects

Prairie Crossing, best known for being a conservation community, is located in Grayslake, Illinois. This community was created to combine preservation of open land, provide residents with easy commuting by public rail, and acquire responsible development practices. It is a nationally recognized example of how one plans a community to improve the environment and bring about support for a healthier way of life.

Residents at Prairie Crossing can enjoy a variety of social events and activities such as gatherings at the historic Byron Colby Barn, gazebo concerts, farm markets, and various summer and winter activities. Amenities within the community include the Prairie Crossing Farm that sells organic produce, honey, and eggs and the Prairie Crossing Charter Public Elementary School. Furthermore, two commuter lines that allow connections to both Chicago and O'Hare International Airport are just walking distance away.

Covering an area of 667 acres 270 hectares, Prairie Crossing provides a total of 359 single-family homes along with 36 condominiums. The new condominiums are two and three bedroom one-level residences that surround the town square and are within walking distance to shops and the public train. The single-family homes were designed with considerable attention to the existing old Lake County homes that are located in nearby towns and farms. The design team created new designs based on their research of Midwestern Vernacular architectural traditions. About 25 single-family homes were included in the initial master plan with each house devoted to the historical context of the site while being 50 percent more energy efficient compared to contemporary homes in the Chicago area.

976

One of the unique aspects of the community is land preservation. Over 60 percent of the 667-acre (270-hectare) site is a protected open area, which is home to local wildlife and used by the public. These areas include numerous trails, farm fields, pastures, lakes, ponds, native prairies, and wetlands. Residents use these trails for pedestrian activities, biking, skiing, horseback riding, and watching wildlife. The colors of the homes also provide a connection with the environment and landscape by using earth tones and warm colors. In addition to land preservation, historical buildings were also preserved for current use—such as the Byron Colby Barn built in 1885 that was renovated in 1996. The site is designed to filter stormwater runoff through the prairies and wetlands into Lake Aldo Leopold where residents can swim, canoe, kayak, and sail during the summer months.

GENERAL DEVELOPMENT PLAN

生态现代公寓
Eco Modern Flats

Fayetteville, Arkansas, United States
Modus Studio

The Eco Modern Flats project is defined as a sustainable living community and is located near the University of Arkansas, Dickson Street Entertainment District and bustling Downtown Fayetteville. It also includes the renovation of an existing 96-unit apartment complex that earned it the first LEED for Multifamily Homes Platinum project in the State of Arkansas.

The renovation included the task of transforming the four existing apartment buildings made of precast concrete and cinder block that were constructed between 1968 and 1972. These materials were unsuitable in terms of thermal comfort, air quality, and aesthetic appeal. Another issue with the existing structure was the way that the units created a social detachment from the site's amenities. To tackle these two issues, the architect redesigned the living systems of each unit and integrated a new visual flow. The existing topography and lost open spaces between the buildings were re-designed and utilized to create a variety of public courtyards, private terraces, patios, and rooftop decks. This new design both created connections with the existing landscape and also meets the needs of occupants by augmenting communal experiences.

To work around the existing precast concrete and split-face block structure, the design re-works the original spatial organization of each unit and introduces a variety of new materials to meet the budget target. Steel was used for durability while cedar was used for natural warmth to create a pleasing new aesthetic. Interior finishes were also redesigned and included materials such as quality wood millwork, concrete countertops, and polished concrete floors to give residents durable, clean, and sustainable finishes while at the same time introducing a new color palette.

The once "boxed-up" units were transformed into a spacious dwelling due to the introduction of large windows, sliding patio doors, and open-living spaces with built-in multi-functional storage and work spaces. Cedar panels were used to shape new, private terrace spaces and balconies. Multiple staircases lead from the third-floor units to a private rooftop and larger stair leads from the pool to the public rooftop deck. This circulation provides residents with views of the university, city, and the surrounding mountains—views that were previously impossible.

The outdoor space was essential for defining and integrating the term "community" in this project. The redesigned landscape created new public spaces for outdoor living, socializing and relaxing. These, in turn, created a new relationship between the buildings, residents and nature. Serpentine retaining walls were included in the design to address the issue of exterior erosion on the sloping site and allow new patio areas for ground-level units to be utilized. New spaces for public amenities were integrated into the former laundry facilities—creating spaces for leasing offices, restrooms, and a public saltwater pool surrounded by vegetation. Further sustainable solutions include a lush vegetated steel-cable trellis system and freestanding water cisterns that were fabricated from steel culverts to provide additional function, minimize costs, and augment the place's visual appeal.

石楠木教堂社区

Briar Chapel

Chatham County, North Carolina, United States
Cline Design

Briar Chapel is a green community located on 1600 acres (647 hectares) of North Chatham County woodlands in North Carolina, and was designed with nature preservation in mind. It includes 24 miles (38.62 kilometers) of planned biking and hiking trails, plus various community parks adjacent to the recreational centers and neighborhoods. Efforts to protect the natural landscape, water quality, and trees have preserved approximately 900 acres (364 hectares) of open space by installing erosion control and stormwater management systems.

Located approximately 5 miles (8 kilometers) south of Chapel Hill and planned under Chatham County's new Neighborhood Model Ordinance, Briar Chapel is known to be one of the largest green communities in the region and has strict green-home construction standards that all homebuilders must meet. These guidelines include use of Energy Star utilities and Healthy Built Homes standards. Efforts to preserve the steep slopes, native wildlife, indigenous plant species, and water resources have also taken place. The planned layout provides approximately 2000 units in a wide range of product types, organized together into a traditional neighborhood style, including a centrally located village center.

Amenities include a center with recreational facilities, community parks, various trails, schools, village center, historical preservation sites, and wildlife observation and preservation. As its centerpiece, the community center includes designated areas for social gatherings, educational uses, indoor and outdoor fitness, and administrative roles. This place also includes a large pool complex, formal event lawn spaces, sporting courts, shaded structures, and allows parking for more than 100 vehicles. The LEED Silver certified community center also exhibits many strategies for sustainable design and planning, and acts much like a themed center point for the entire development.

第二章：移动思维

MOVING AROUND

Designing mobility and networks with a suitable integration of streets, bike, and pedestrian paths, and reducing the use of private vehicles is central to the planning of smart communities. This chapter discusses and illustrates the most appropriate road, parking, and shared street patterns for such developments, their effects on circulation, and the resulting curb appeal.

Reducing the use of fuel-consuming, polluting cars is critical for lowering emissions rates and reversing global warming trends. The goal is to design appropriate streets and parking spaces, while safely integrating pedestrian and bike paths. Planning a well-connected community by considering these aspects contributes to establishing well-integrated places based on sound ecological, economic and social principles (Bartlett 2009).

Well-connected communities are beneficial to the residents' health as well. A United Nations report suggests that a significant number of deaths worldwide are caused by air pollution and lack of exercise (Hood 2011). In addition, a 2005 survey by the Canadian Heart and Stroke Foundation found that people who live in suburbs, smaller towns, and rural areas are often at a higher risk of heart disease and stroke than their city dwelling counterparts.

The outcome of physical inactivity is the rapid increase in obesity and its associated health risks. It was demonstrated that each additional 0.62 mile (1 kilometer) walked per day reduces a person's likelihood of becoming obese by nearly 5 percent, while each hour per day spent in a car increases the likelihood of becoming obese by 6 percent. The key attributes of communities that consider mobility reflect those needed to improve residents' health, namely the integration of pedestrian and cycling pathways and network connectivity.

Emissions from motor vehicles are known to produce a number of air pollutants that pose public health risks. When people breathe in polluted air, their respiratory system is greatly affected resulting in inflammation of the lungs. Individual reactions to air pollutants depend on the type of pollutant, the degree of exposure, and the individual's health status (Road Traffic and Air Pollution 2011). Research demonstrates that it is crucial, therefore, to integrate public transit when designing a community to reduce these pollutants. A study by First Group plc found that a bus emits 3.13 ounces (89 grams) of CO_2 per 0.62 miles (1 kilometer) traveled while an average size car emits 4.69 ounces (133 grams) per 0.62 mile (1 kilometer). In sum, a vehicle emits 50 percent more CO_2 per mile than the bus (R. Angel 2013).

The planning for mobility begins by accounting for regional accessibility. The proximity of a site to other major regional destinations, such as employment and amenities, will result in a shorter commuting distances. A research study comparing two automobile dependent neighborhoods in metropolitan Nashville, Tennessee found that the residents of the neighborhood closest to the city drove 25 percent less than those who lived in further neighborhoods because they had access to public transit (Piedmont-Palladino and Mennel 2009).

For the most successful creation of a well-connected neighborhood with lower levels of carbon dioxide emissions, some basic design principles need to be considered. Prior to designing streets, planners should preferably select a site where a single or multiple transit lines are already available. On average, a bus stop should be located no more than 3000 feet (1000 meters) away from a home. The right mix of land use, efficient public transportation, and the integration of streets and paths will foster greater walkability (Piedmont-Palladino and Mennel 2009). The order of priority should be given first to pedestrians and cyclists, then to public transit, and finally to private vehicles.

In some cities like Copenhagen or Amsterdam, pedestrian-only streets within a high-density community offer shopping opportunities that are solely connected by pedestrian paths (Rodriguez 2013). Therefore, a possible design option is to create large-scale pedestrian-only zones in central business districts that will inevitably eliminate vehicular traffic. Visitors would be encouraged to walk since it would be safer. Streets would then be regarded as public spaces for social interactions as well as network connectivity. Trees and benches should also be included in the design of the path walkways to attract pedestrians.

Using a walkability index based on community density, mixed-land use, and street connectivity, strategies for Metropolitan Atlanta's Regional Transportation and Air Quality (SMARTRAQ) rated sites near the City of Atlanta, Georgia, and concluded that the travel patterns of residents in the least walkable neighborhoods resulted in about 20 percent higher CO_2 emissions per capita than the travel by those living in the most walkable neighborhoods. That comes to about 2000 more grams of CO_2 per person each weekday (Piedmont-Palladino and Mennel 2009). The speed of cars and the carbon footprint can be reduced further by paving streets with bricks instead of asphalt. The proper integration of the streets in order to enhance pedestrians' mobility and safety will greatly affect sustainability.

Parking spaces also need to be carefully thought through and sited to reduce negative impact on the environment. The space allotted for parking should be minimized to preserve the surrounding flora and fauna. Commonly, paving can consume up to 50 percent of an entire development, therefore, reducing the amount of area set aside for parking spaces and roads would increase the space allotted to greenery. As parking lots have become a dominant feature of urban and suburban landscape, their environmental impact has also become increasingly apparent (Green Parking Lot Resource Guide 2008). They are usually built of a mix of asphalt, concrete, sand, and gravel that absorb heat but do not allow the filtration of rainwater into the soil. The water travelling on the pavement's surface accumulates chemicals such as petroleum, fertilizers, pesticides and other pollutants, thereby negatively affecting the environment (Green Parking Lot Resource Guide 2008).

The planning of a community needs to begin with a consideration of parking alternatives by integrating them with other community features, such as buildings and open spaces. In general, private driveways for each home should be avoided. Shared parking spots have positive outcomes on the environment and also on the residents of the community. Furthermore, they reduce the cost of a home. One parking space per unit increases the cost by 12.5 percent and two parking spaces increases the cost by about 25 percent. Residents find that shared parking lots have helped create vibrant places as they become social meeting places, while protecting environmental quality and still providing for necessary vehicle storage (Sustainable Communities 2012).

Downsview Park in Toronto, Canada, features parking lots with different surfaces that are more like parks. The urban heat-island effect and stormwater are managed to mitigate potential environmental damage. The large areas of parking are broken up into courts and separated by pedestrian pathways. These areas are also surrounded by planting, which diminishes their visual impact. The pavement is made up of a permeable material that absorbs rainwater, which is then filtered by the soil (Sustainable Community Development Guidelines 2007). In sum, the space allotted for a car should consider the environment and the needs of a community.

The innovative planning of parking can also encourage utilization of public transit. In Portland, Oregon, the La Salle Apartments have reduced their parking space to 1.8 spaces per unit and they provide pass allowances to residents. This has resulted in a large increase in transit ridership among the residents of the apartment building (Sustainable Communities 2012).

Highland's Village Garden in Denver, Colorado, was recognized with the Urban Land Institute Award for Excellence in 2007 as well as the International Economic Development Council in 2006. It is a well-connected sustainable neighborhood that considers mobility. The 23.7-acre (9.59-hectare) community is built on a traditional street grid and features many pedestrian paths (Highland's Village Garden 2013). Pedestrian paths were well integrated for efficient mobility and the neighborhood successfully gives residents the opportunities to live, work, and shop within minutes of their homes, resulting in an environmentally friendly atmosphere and a successful connectivity within the community.

Some European nations place the needs of pedestrians and children ahead of cars (Tumlin 2012). Drivers need to drive with great care because the pedestrians are sharing the street with them. In the Netherlands, streets named Woonerf (meaning "streets for living") are shared among pedestrians, bicycles, public transit vehicles, and automobiles. These shared streets are safer for pedestrians since the cars travel at slow speeds. Streets with a width of 20 feet (6 meters) have been observed to benefit both pedestrians and drivers since the drivers proceed more slowly.

Public transit is an ecologically sound way of traveling. In Zurich, the traffic lights allow trams to proceed before private vehicles. The implementation of this transit priority program has improved the efficiency of the public transportation system and created an extremely attractive transit network at a fraction of what it would have cost to build a capital-intensive underground metro system. Cities including Boston, San Francisco, and Brussels have built light-rail transit systems (LRT) that were significantly improved by giving them priority programs similar to Zurich's (Nash 2010). A community sited near public transit such as that in Zurich is a great advantage for residents as they are strongly encouraged to use it. Moreover, residents of communities who are sited further away have the opportunity to bike to the nearest public transit station and leave their bikes in a safe and convenient parking area. According to Siemens' Green City Index, in Zurich 48 percent of the working population walks or bikes to work and an additional 21 percent take public transit. The successful integration of pedestrian pathways and public transit within communities results in positive effects on both the environment and the economy of the neighborhood, as well as the inevitable health benefits.

In summary, in the design process of a sustainable community careful attention should be paid to the streets, paths, and parking spots, and to the integration of public transportation as they affect the overall cost of the community and greatly reduce the carbon footprint.

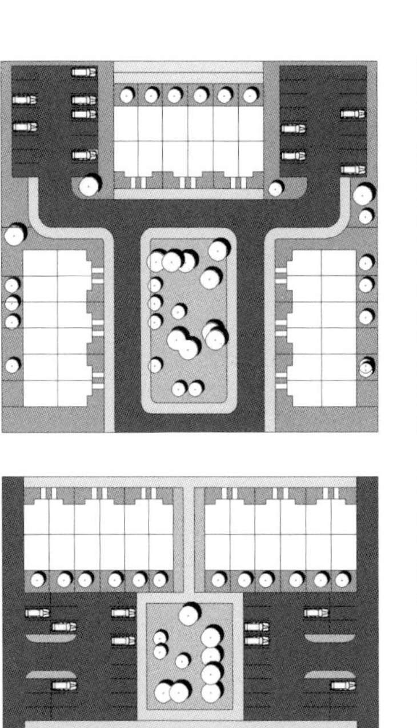
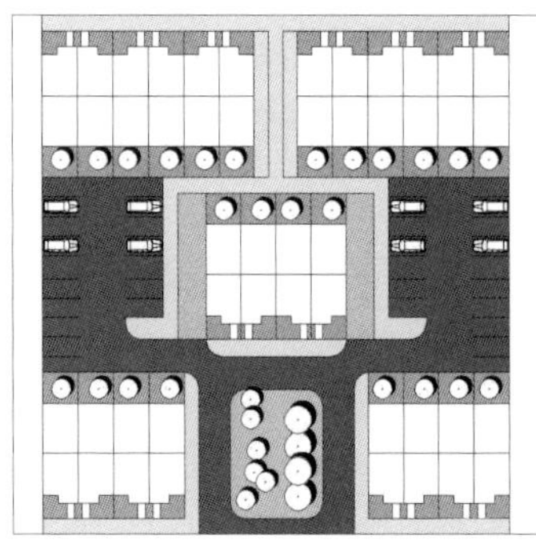
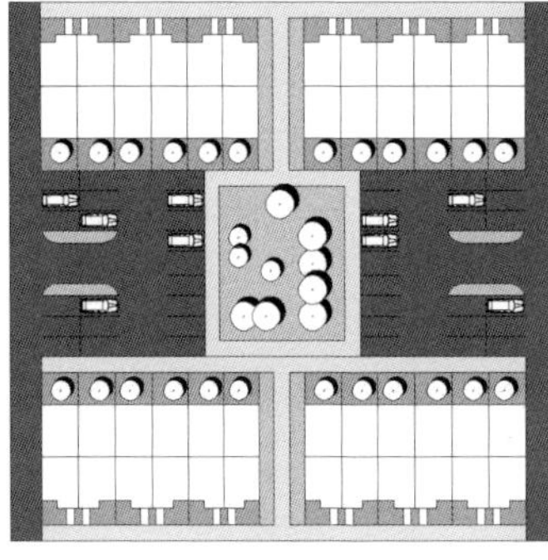
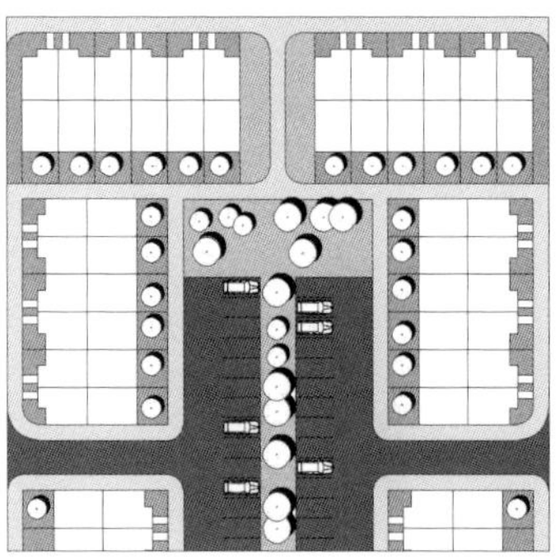
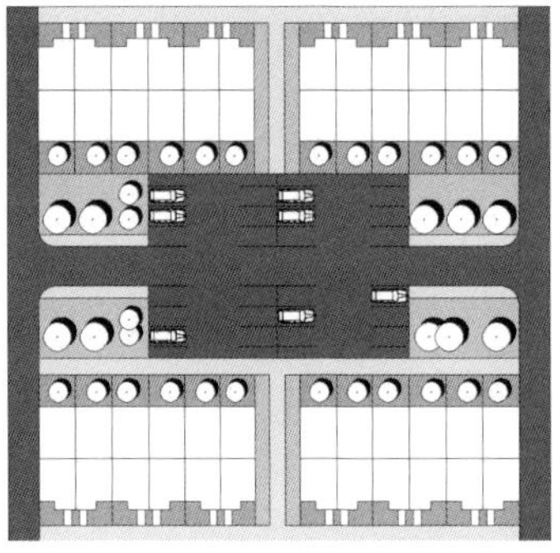
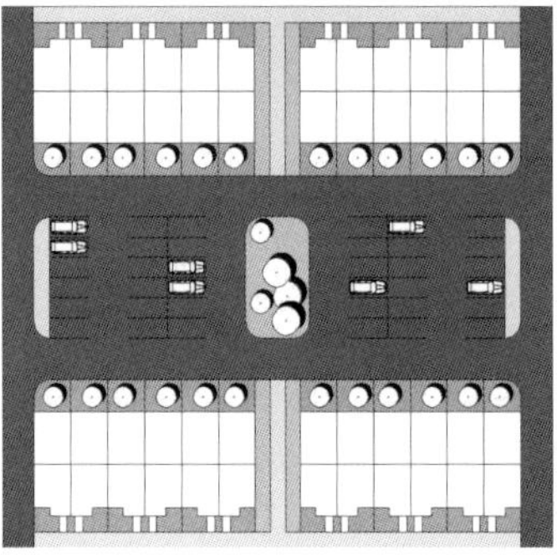

ALTERNATIVE LANE PARKING ARRANGEMENTS IN MID-DENSITY NEIGHBORHOODS

8号公寓

8 House

Copenhagen, Denmark

BIG (Bjarke Ingels Group)

Covering 15 acres (6 hectares) in the maturing neighborhood of Orestad near Copenhagen, this design allows for ample active mobility. It provides inhabitants with the opportunity to bike from street level up to the tenth-floor penthouses of the apartment units with terraced gardens. The complex, which is shaped like the number eight, includes buildings of three different types of housing, retail and office space. Rather than exhibiting a more traditional style of neighborhood, 8 House stacks all the main features of an energetic urban community into horizontal layers, each connected with a network of walking and cycling paths. These main attributes create a three-dimensional urban neighborhood where a more suburban lifestyle intertwines with the energy of a traditional city, allowing business and housing to mix.

With functional architecture in mind, the shape of 8 House creates two enclosed interior courtyards, each separated by the center of its "number eight" shape, which contain communal facilities. At this center lies a 30-foot (9-meter) pathway that allows residents to easily circulate from the western park area to the eastern canals. The various building functions are distributed horizontally instead of being separated into independent structures. Residential units are placed on the top levels allowing views, sunlight, and fresh air, while the buildings designated for business are at the base of complex merging the commercial with street life. The commercial functions include cafés, day-care centers, retail spaces, and offices.

The 476 dwelling units offer three different types of accommodation such as apartments of various sizes, top-floor penthouses, and townhouses, creating a mixture of suburban serenity and urban vibrancy.

café

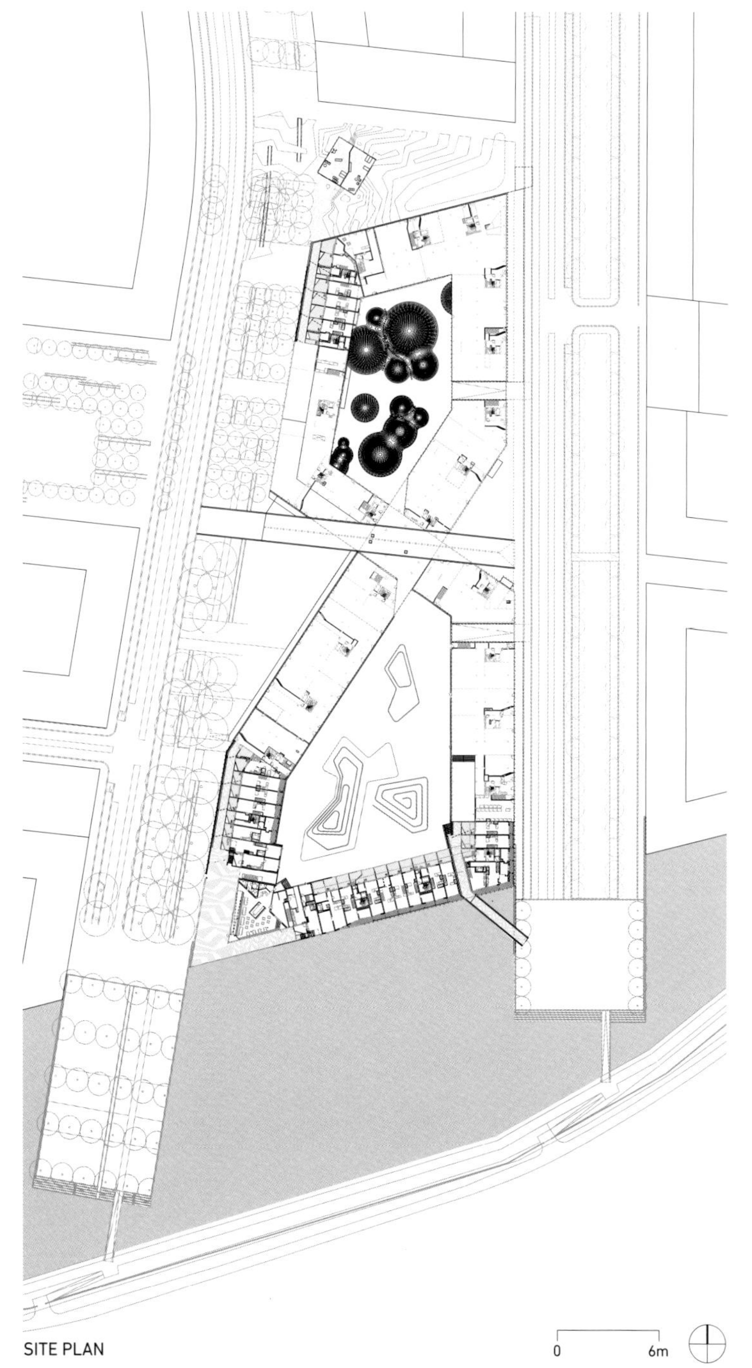

SITE PLAN

The network of public paths, stretching from street level to the penthouse apartment complexes, allows the inhabitants to travel by foot or by bike from the ground floor up to the tenth floor, circulating around the townhouses, which have terraced gardens on the urban perimeter of the site. This network is designed as a continuous open-air ramp, along with stairs and elevators, providing accessibility to the townhouses and penthouses. This alleyway between 150 townhouses fills the site, where it intertwines from ground level to the top and back down to street level again—allowing bursts of social life and interactions to take place. Using the size of 8 House as an advantage, the design exhibits differences in height among the units, creating a unique sense of community.

Steering away from the more traditional style of urban organization, this neighborhood can be considered a three-dimensional urban complex, merging private and social activities. The unique shape of the building allows for this mixing to occur.

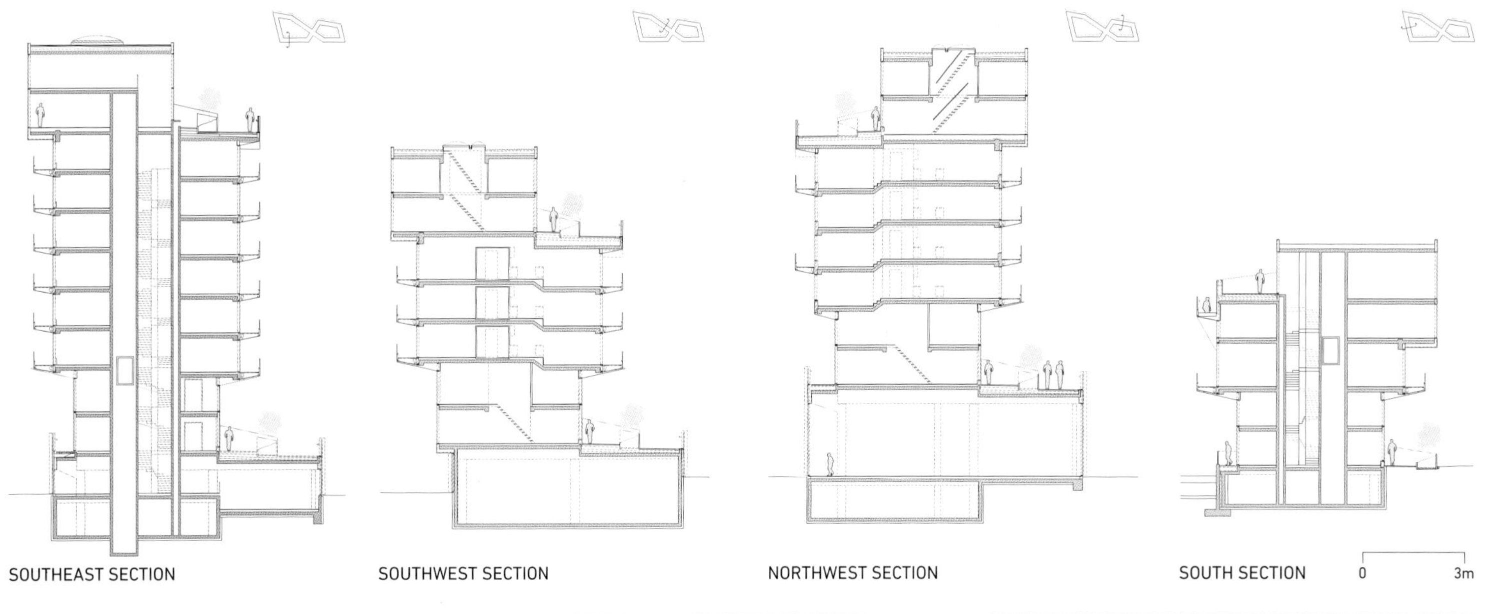
SOUTHEAST SECTION
SOUTHWEST SECTION
NORTHWEST SECTION
SOUTH SECTION
0
3m

奥伦科车站社区

Orenco Station

Hillsboro, Oregon, United States

Costa Pacific Communities, PacTrust

Orenco Station is defined as a modern mixed-use neighborhood located in the city of Hillsboro, Oregon. Known for its combination of accessibility and livability, it is also recognized for its understanding of the health benefits of shopping locally for produce and services. Encompassing a total area of 190 acres (76.9 hectares), this transit-oriented community is conveniently centered around the Orenco Station Light Rail MAX stop in Hillsboro. Orenco Station is also a model of "new urban" design for the future, with various types of housing, an urban town center, and never-ending parks and open green spaces.

Originally designed and re-designed by Costa Pacific Communities as a zone for high-density housing, it was then arranged as a transit-oriented community with varying densities around a town center. The community was also built adjacent to the Intel Ronler Acres electronics plant where most Orenco Station residents are employed. Based on initial research and surveys conducted by Costa Pacific Communities, the design of the houses emulates the historic Portland suburbs with their attention to craftsmanship, bungalow-style architecture, picturesque rose gardens, and neighborhood-oriented shops. This architectural design also had to appeal to the target demographic of young professional couples and singles. The total of 449 housing units comprises townhouses, condominiums, lofts, and cottages.

The produce and services offered to residents supports the local economy in terms of income, employment opportunities, and tax benefits. Other amenities within this large neighborhood include schools, office spaces, restaurants, retail shops, markets, medical centers, and a farmers market, all of which are easily accessible.

NW CORNELL RD
OPEN

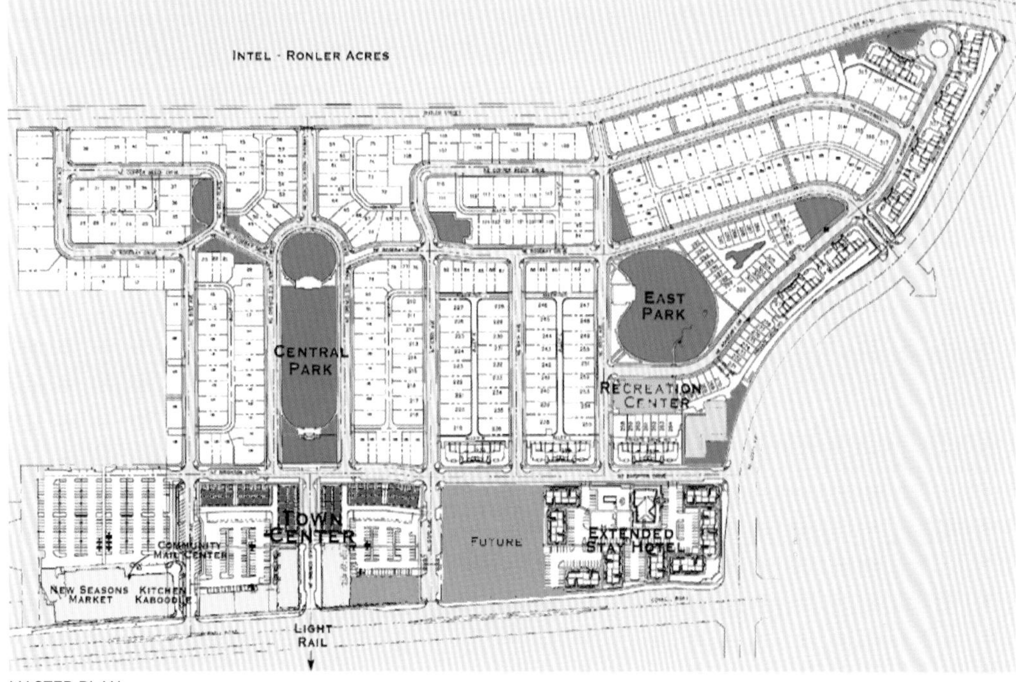

MASTER PLAN

The master plan for Orenco Station was organized around a network of pedestrian pathways extending north from the Westside Light Rail System, through the residences, and finally to the town center. On average, it takes around 45 minutes to reach Portland and 8 minutes to get to Hillsboro by train. The organization of the community was designed to create higher density centers with easy access to goods and services.

假日近邻社区

Holiday Neighborhood

Boulder, Colorado, United States

Barrett Studio Architects, Boulder Housing Partners

Barrett Studio Architects, in conjunction with Boulder Housing Partners, created the master plan for the Holiday Neighborhood in Boulder, Colorado. It is a mixed-use development in terms of income and types of housing units offered.

Built on the remains of a historic drive-in theater and encompassing a total area of 27 acres (10.92 hectares), the master plan of Holiday Neighborhood is focused on a system of pedestrian-friendly open spaces that define the central core of this community. This core then branches out to different indoor and outdoor spaces including a total of 333 different housing units, offices, live/work spaces, retail shops, a massive green open public space, various green pocket parks, and public communal gardens. Together they combine the livability of an urban neighborhood with modern, sustainable green techniques that are energy efficient, cost effective, and environmentally manageable.

The master plan was organized to include transportation pathways with a greenway and park system, a large open green space, and public gardens. The blocks of houses include shared parking in the back alleyways making way for pedestrian circulation. There is even a pedestrian mall that extends from the park to the communal gardens and through a network of live/work units. Commercial and mixed-use spaces line the perimeter and central axis of Holiday Neighborhood. In addition, the master plan includes more shared green spaces and courtyards as opposed to private gardens, to increase spontaneous social interactions. In the neighborhood, bikes, public transit, and pedestrian circulation is highly encouraged over automobile routes through creating easy access to amenities and employment spaces as well as providing residents with inclusive bus passes. Furthermore, the public transit stations are placed so that all residents of each zone have easy access and the city bike pathways extend along all major roads of the community.

Of the 333 housing units, half of which are affordable, ten units are set aside for people in the process of transitioning from chronic homelessness, three units for homeless families in emergency situations, and another ten are for clients of the Boulder County Mental Health Center. This is a unique method of truly providing

all individuals with a chance of purchasing their own home or unit while giving them the tools to do so in a safe environment. The designers along with a diverse team of architects, developers, and builders, followed strict guidelines for each building zone based on both qualitative and quantitative standards that dictated siting, massing, parking, lighting, landscape, architectural elements and materials—making this community a leading example for future sustainable communities.

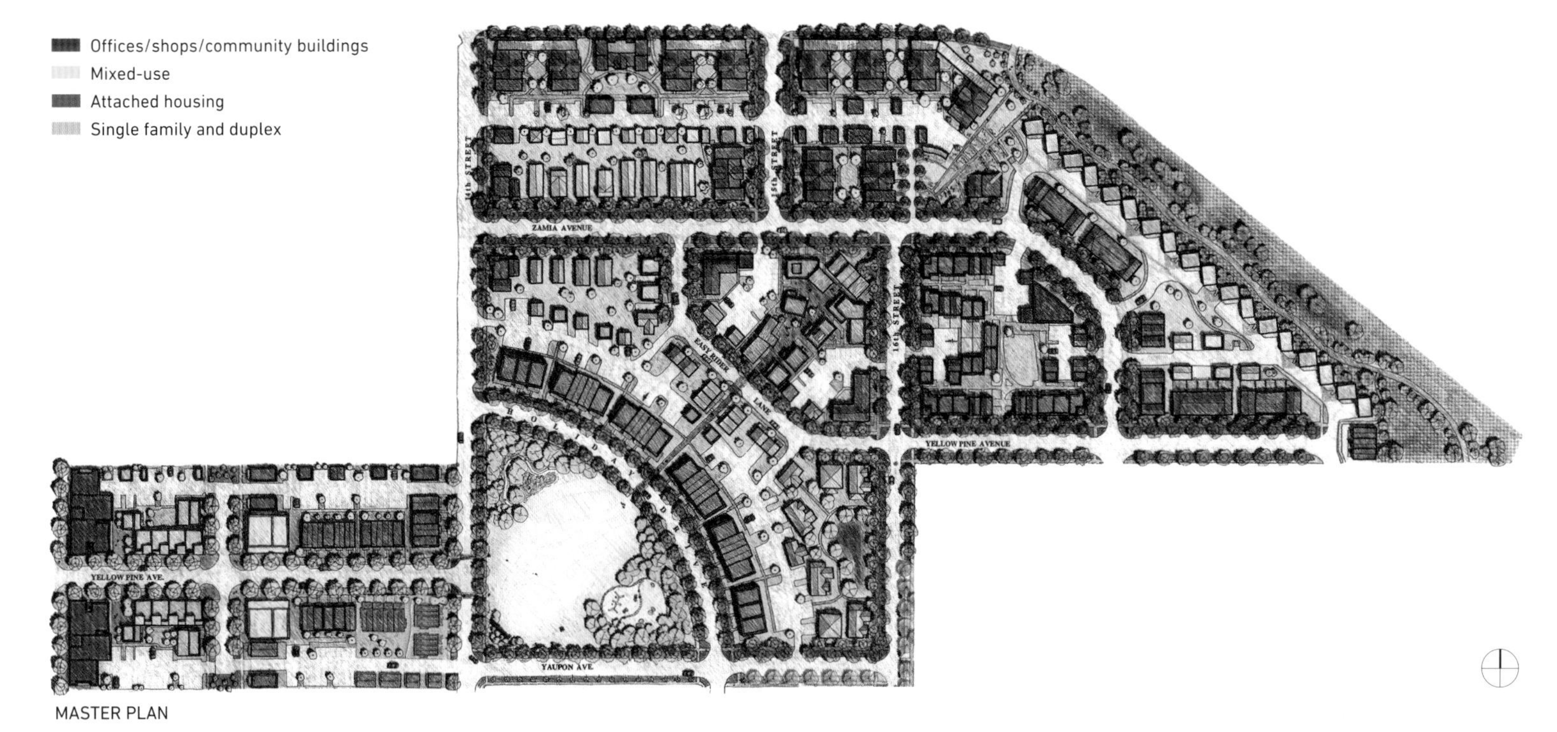

Offices/shops/community buildings
Mixed-use
Attached housing
Single family and duplex
ZAMIA AVENUE
YELLOW PINE AVENUE
YAUPON AVE.

MASTER PLAN

Broadway
SPEED LIMIT 25

昆次伏广场社区

Kuntsevo Plaza

Moscow, Russia

The Jerde Partnership

Designed by The Jerde Partnership, Kuntsevo Plaza provides a new dynamic area devoted to public activity and distinct commercial functions in central Moscow. This pedestrian-oriented plaza contains vibrant leisure, shopping, business, and residential complexes. Together, it reaffirms the urban fabric of the historical Kuntsevo district, while creating a new landmark for the city.

Kuntsevo Plaza is located at the intersection of pedestrian and transportation routes, between Rublevskoe and Mozhaiskoye highways, occupying an area of 60 acres (24 hectares). The design's main objective is to establish a connection with its surroundings. This includes a transit line that contains a large public plaza at the main entrance closest to the Molodezhnaya Metro Station, and a number of access points from the surrounding streets, which all take advantage of the changing topography of the site.

The first of its kind in Moscow, the complex mixes residential, retail, entertainment, offices, and park spaces, paving the way for future sustainable development. More specifically, it includes universities, city administration offices, medical facilities, historical sites, and residential amenities, allowing the residents, business people, and tourists to be under one roof. With all these experiences in one place, the unhindered flow of pedestrian traffic is achieved. With nature serving as a contextual element in the design, the experience of the four seasons, the extension of green spaces, and vegetation are celebrated. This also helps guide the natural course of the pedestrian streets, plazas, courtyards, and indoor/ outdoor green park terraces that foster social interactions, encourage healthy lifestyles, and stimulate new discoveries.

STARBUCKS COFFEE
LUSH
СВЕЖАЯ КОСМЕТИКА РУЧНОЙ РАБОТЫ
АШАН
АШАН
АШАН

Above the pedestrian pathways is a multi-level commercial street that comprises international shopping areas, restaurants, and entertainment venues. In addition, three office buildings and two high-rise residential towers, occupying 50 acres (20 hectares) and overlooking the public activity below, permit unity and social spontaneity among tourists and residents. The inclusion of escalators and large staircases guide visitors to the public park terraces above, each of which has a different public function. This unique pedestrian system controls circulation and distinguishes public from private functions. Ten distinct levels with a specific function for each encompass the plaza, including four levels of underground parking for 2000 cars.

Another central aspect of the plaza is the urban connection between the Kuntsevo neighborhood and the local residents. For example, the road and pedestrian connections between the metro stations and the State Aviation Technology University are directed through the existing urban patterns around the site—thereby preserving the former urban organization. This circulation strategy creates an unhindered flow from the nearby metro stations, through the plaza and to the residential neighborhoods for commuters as well as visitors.

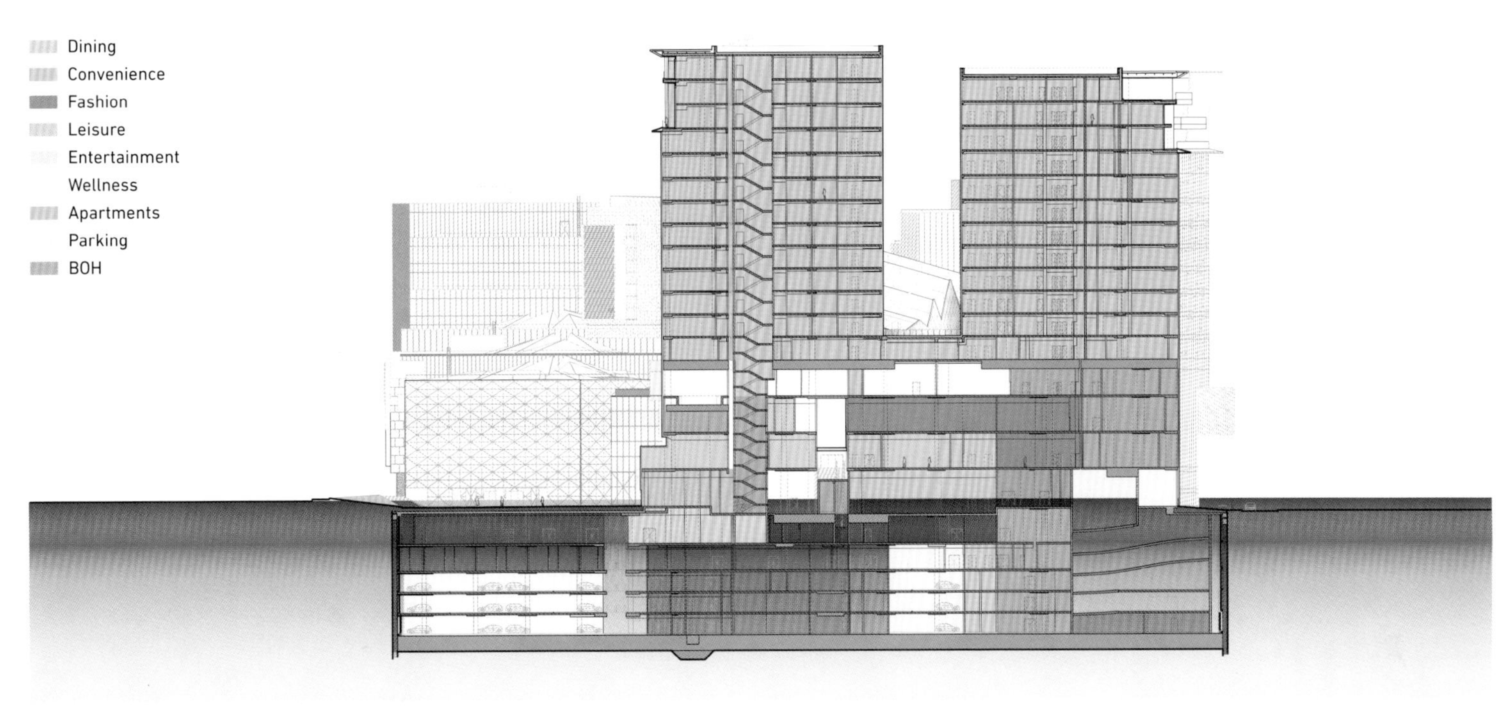

SOUTH-EAST SECTION

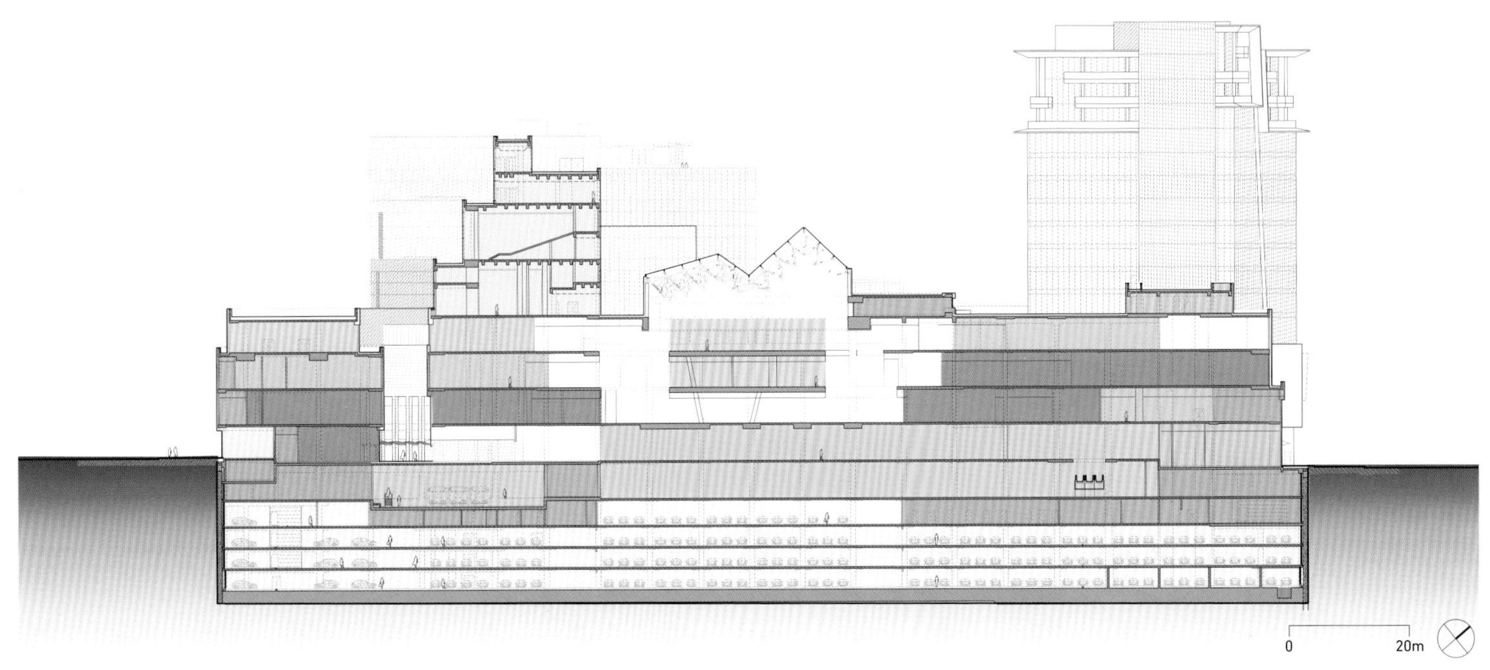

SOUTH-WEST SECTION

第三章：基础设施与住宅的混合思维

MIXING AMENITIES AND RESIDENCES

3

The common tendency when planning low-density neighborhoods is to segregate residential and non-residential activities. Such practices encourage reliance on private cars and construction of an extensive road network. Commerce and other communal functions, such as libraries or medical clinics, can be placed in a central location to which pedestrians' paths lead. Alternatively, if these functions are located near the perimeters of a development they can also be accessed by residents of neighboring communities, which further supports these uses economically. This chapter focuses on principles for integrating non-residential land uses into neighborhoods.

When planning a residential development, the planner must consider and allocate space to the amenities that will be needed by the inhabitants. Proximity and accessibility to basic services such as grocery stores, schools, childcare centers, and medical clinics are essential. Locating them within reach of every home will make them accessible and usable by pedestrians and cyclists. Moreover, such establishments should be designed to be adaptable for when changes in economic or functional realities take place. Mixing amenities with residences is therefore key to successfully creating well-functioning neighborhoods (Bartlett 2009).

Currently, low-density suburban developments often do not have enough inhabitants to economically support the integration of amenities. Families must drive to reach them resulting in additional household expense and negative environmental consequences. When housing is located in walkable, transit-oriented communities, the percentage of household income spent on transportation drops from 30 percent to approximately 9 percent, freeing the remaining income for other uses such as education and saving for old age (Piedmont-Palladino and Mennel 2009). Developments in or near an existing metropolitan area will also generate the most economic benefit to the municipalities since fewer roads will be needed

Planning and developing new communities involves far more than solely building homes and roads. When a mixed neighborhood is planned to include public gathering spaces, a sense of belonging is established. Meeting or being in the company of other people is also beneficial to people's mental health. It is important, therefore, to recognize the role of socially-oriented design in improving the wellbeing of the residents (Bartlett 2009).

The mix of land uses is governed by zoning regulations and by-laws that dictate how and where buildings and other civic functions such as parks can be located, the types of buildings permitted, lot sizes, parking requirements, heights, and setbacks from the streets to name a few. These regulations also specify the preferred types of amenities; for example, forbidding constructions of a polluting factory near dwellings. Zoning laws for which amenities and how many of them are allowed to mix with residences commonly differ from one community to another since the density and the population would also be different.

Generally, higher density communities will have an increased mix of amenities. The Fehr and Peers synthesis of national research on neighborhood form and travel behavior concluded that independent of other factors, a doubling of households per acre would lead to a 5 percent reduction per household in miles driven and a 4 percent reduction in vehicle trips, compared to a base case. Furthermore, the firm states that, independent of other factors, mixing commercial and residential buildings so as to double the ratio of jobs to housing would also result in a 5 percent reduction in miles driven, along with 6 percent reduction in vehicle trips. Doubling the number of households per acre plus mixing commercial and residential buildings would lead to a 10 percent reduction in miles traveled and the lowering of the associated carbon emissions, on top of whatever benefits may be realized by the site's general location (Piedmont-Palladino and Mennel 2009).

The relationship between higher density and mixed-land uses in encouraging walkability is well established. A comprehensive multi-year study of land use, travel behavior, and health in Atlanta by Georgia Tech University in the United States found that people who live in neighborhoods with a mix of shops and businesses within easy walking distance are 7 percent less likely to be obese and that their relative risk of obesity decreases by 35 percent. A person living in a compact community with nearby shops and services is expected to weigh 10 pounds (4.53 kilograms) less than a person living in a low-density residential-only cul-de-sac subdivision (Piedmont-Palladino and Mennel 2009). Therefore, the mix of amenities in a community that is within walking distance of homes has a substantial impact on the health of residents.

Decisions as to which amenities should be included are based on the prospective number of users of such facilities and are strictly an economic decision. To work out the proper ratio of residential to commercial buildings, a Master Plan (MP) value is calculated. The MP is determined based on the types of residences and the make-up of the population. For example, the MP value for the year 2003 intended for mixed-use developments (i.e. commercial and residential zones) is determined based on the percentage of the mix of uses allowed for such zones. For different environments and circumstances, the MP 2003 value can be computed solely based on the use intended for the zone (determining the MP 2003 Value 2007). Among the services that should be mixed with residences are childcare facilities, schools, public transportation, groceries, healthcare clinics, and green open spaces.

Mixed communities also help to preserve more of a site's natural features and reduce forest clearance. Creating green open spaces for passive or active recreation use is beneficial for inhabitants of all ages. Furthermore, public transit can be integrated according to the lay of the land. The design of the streets can follow the site's curves and slopes, and bus stops can be located near trees to provide shade during summer.

The amenities of a community should preferably be central so that most residents will be able to reach them on foot. Homes should also be linked to the amenities with pedestrian and cycling paths to further encourage residents to walk or bike to them. In addition, as a matter of sustainability, selecting a site with, or adjacent to, existing amenities rather than building new ones is desirable. Conceiving a place that will leave land for such facilities to be added in the future is also advisable.

Buckman Heights is a mixed-land use community which was developed near downtown Portland, Oregon. It has apartment buildings adjacent to amenities and public transportation. The site is near Portland's central city Lloyd District and on the edge of a light-industrial area, used for decades as a car dealership. Despite an overheated real estate market, the 3.7-acre (1.5-hectare) site had been on sale for well over a year without attracting any interest. A development firm named Prendergast and Associates saw an opportunity to build housing on the site, given its prime location—nine blocks from light rail, within five blocks of four high-frequency bus lines, and surrounded by a growing network of bike lanes. Prendergast developed the 2.5 acres (1 hectare) of vacant parking lots into sites for 274 units of housing (8 unit townhouse project), a 144-unit mixed-income apartment building, and a 122-unit apartment building with a small retail space. Creative parking strategies helped to keep development costs low. The minimum parking allocation required by the city is 0.5 spaces per unit, that was reduced further to 0.4 spaces per unit. Since the neighborhood was close to public transit and bicycle paths, the developers were able to reduce the parking requirements by introducing generous sidewalks and promoting bicycle facilities, on-street parking, shared off-site parking, unbundled parking costs, and car sharing (Sustainable Communities 2012).

The developer also introduced a shopping center, creating a mix of public and private environments. The intention was to add entertainment and cultural attractions to the residences and to enable their use. The city went on to establish new zoning for the district to allow arts and entertainment venues and to exempt the developer from standard parking requirements by allowing shared parking in off-site public parking structures (Sustainable Communities 2012). The new zoning permitted outdoor cafés, which made the place more inviting and attractive.

During the design process for mixed-use neighborhoods, careful attention should be paid to location, type of residences, and amenities, as they affect the overall cost and can greatly reduce the carbon footprint. Amenities should be integrated as much as possible within a community while keeping a fair ratio of residents to commercial buildings. Having amenities also plays an important role in the community's social network, as the residents are more likely to have a place to meet.

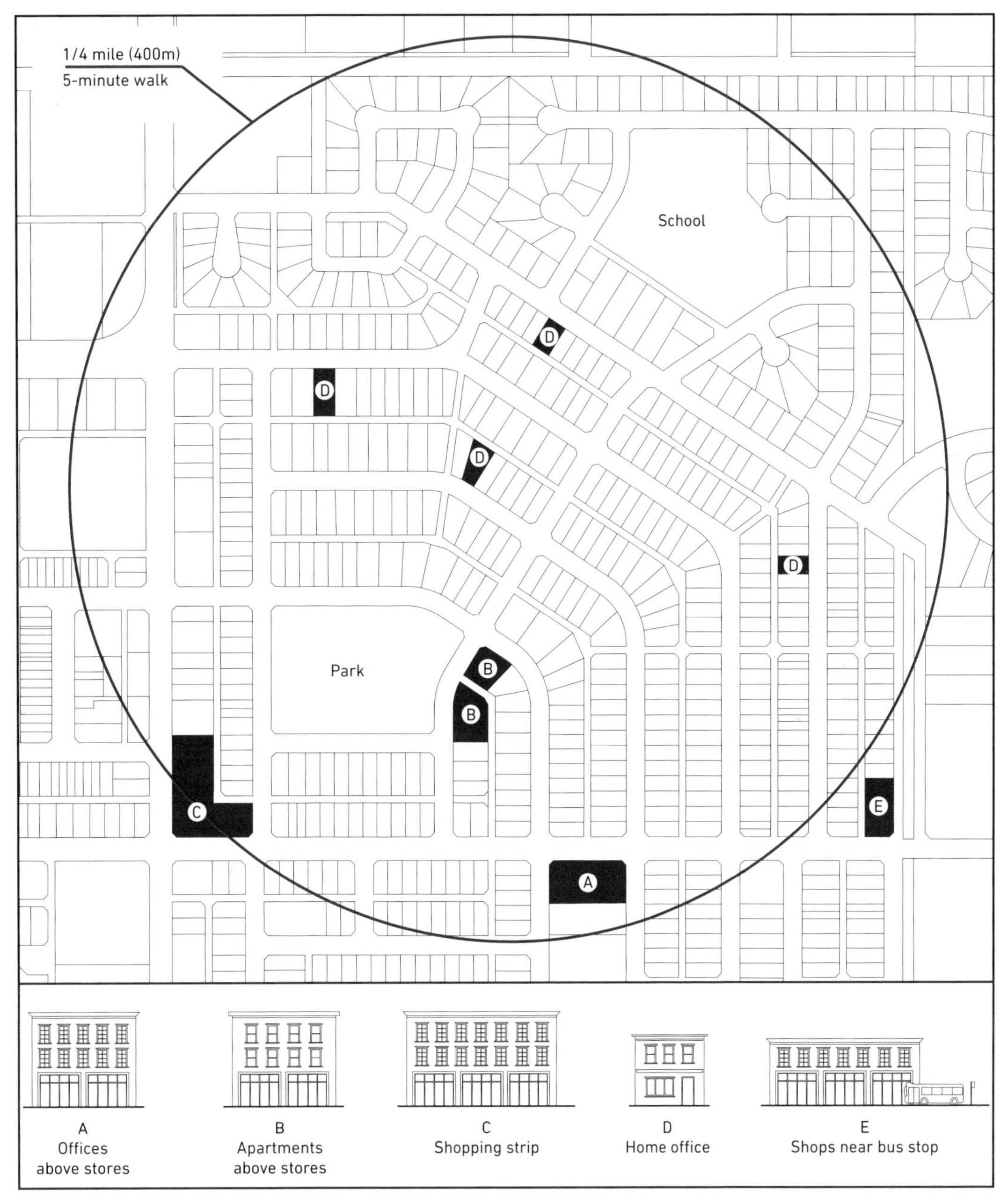

WEAVING RESIDENTIAL AND NON-RESIDENTIAL LAND USES WILL HELP REDUCE RELIANCE ON PRIVATE CARS AND RENDER A PLACE MORE SUSTAINABLE.

前景新城

Prospect New Town

Longmont, Colorado, United States

DPZ CoDESIGN

Prospect New Town is Colorado's first New Urbanist community and is located on a former 80-acre (33-hectare) tree farm, just 10 miles (16 kilometers) northeast of Boulder, Colorado. It is the creation of a pedestrian-friendly community that is at the heart of Prospect's block structure, thoroughfare sections, landscape design, as well as its retail, recreational, and amenity programs. The hope is that residents will take a pleasant, convenient walk to all the many services and destinations rather than drive. This distinctive project, in addition to minimizing its impact on the land, also features narrow, tree lined streets connecting homes to numerous parks and public amenities, shops, and offices. The types of dwellings vary from detached homes, townhouses, courtyard houses, apartments, and loft units. Mature trees have been planted along the town's streets and parks to provide the shade and privacy typical of an established community.

The design of neo-traditional communities such as Prospect New Town offers the opportunity to reduce car use in favor of bicycle and bus transportation. This is possible by careful town planning of roads and locating of amenities. In the early 1990s, developer John Wallace wanted to give new life to his family tree farm and have a community that placed people's needs ahead of cars. He also wanted the place to respect the historic character of the area while incorporating the mature landscape features of the farm. DPS CoDesign master palnned the community in 1994 and continues to work closely with the town founder to apply and refine New Urbanist principles. In addition to careful placement of streets, parks, and amenities, they also provided design guidelines to create a unified architectural style.

Located just southwest of Longmont, Colorado, the surrounding environment has a breathtaking view of snow-capped peaks and farmland. One notices a unique street layout when walking through the town; the roads are aligned to take advantage of the mountain views. The importance of combining the exterior landscape with the streetscape design encourages social interaction among the residents. This combination, along with providing a connection to the public amenities, further distinguishes Prospect New Town from other high-end suburban developments. For example, there are pocket parks in proximity to each house. In addition, a large community park surrounded by shops, a swimming pool, an ice rink, restaurants, and offices can be found in the two-block town center and within a five-minute

TENACITY DR

FOR SALE
TOYOTA

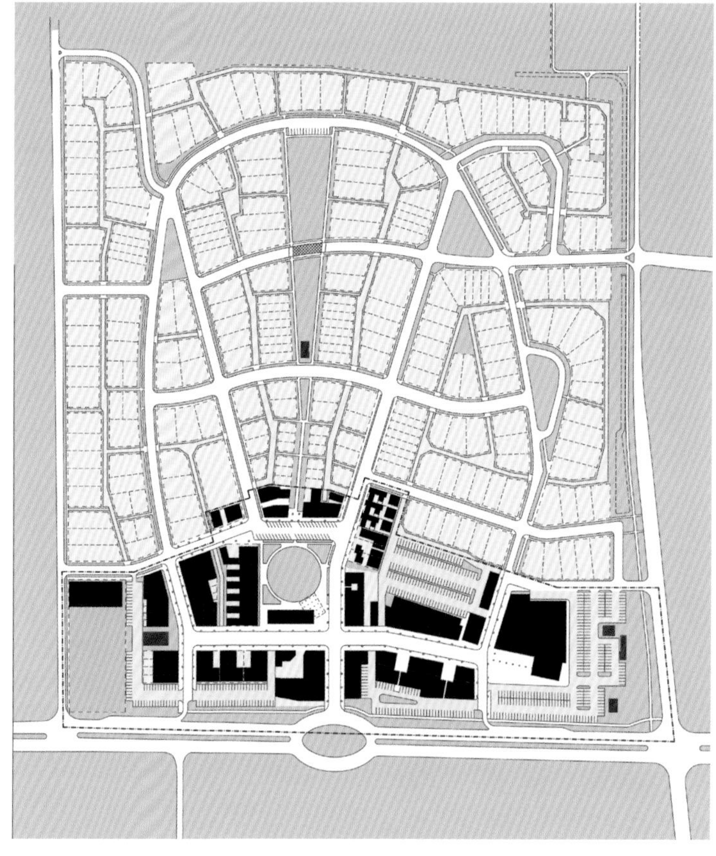

walk from the homes. One of the commercial buildings located in the town center includes offices, a coffee shop, a bookstore, and a post office.

Ornamental landscaping and streetscaping is also incorporated in the design, with benches, custom brick, and concrete-paved narrow streets. Alleys provide a reduction of impermeable areas and runoffs into a stream nearby. Prospect New Town's success is not only measured by housing sales, but also by the recognition it is receiving. Its performance as a community, its effect on public life and social interaction, as well as its support of local goods and services are all rooted in its careful and well-thought planning.

马斯达尔城市开发区

Masdar City Development

Abu Dhabi, United Arab Emirates
Foster + Partners

The design goal for Masdar City was to be a carbon-neutral desert community that also had zero waste. To achieve this, the design by the London-based architectural firm of Foster + Partners combines contemporary technologies with the planning principles of traditional Arab settlements. The 1483-acre (600-hectare) project was a key advance for the government of Abu Dhabi in the development of renewable-energy and clean-technology solutions in a region with an extreme desert climate.

The mixed-use low-rise project fosters a high-density urban structure and includes headquarters for the International Renewable Energy Agency and the Masdar Institute. Just 10.56 miles (17 kilometers) southeast of Abu Dhabi and conveniently located for Abu Dhabi's transportation infrastructure, Masdar City is connected to neighboring communities and the international airport by pre-existing road and railway routes. Due to the hot climate of the Middle East, the project was faced with the challenge of becoming the first modern community worldwide to operate without automobiles at street level.

Since the maximum distance between the residential areas and the nearest public transit stop and amenities is 0.12 miles (200 metres), the city is organized to promote walking. Walking, as a form of transportation, is further encouraged by using various building types and street design techniques to protect pedestrians from the heat.

The public transportation network, on the other hand, is planned to be below street level. Buildings that are placed at the city center only rise to about four to five stories above that level. Since there are no cars at street level, the main circulation can be narrower and buildings can provide even more shade for pedestrians. Therefore, the careful design of streets and buildings can help create comfortable living environments that can reduce the need for air conditioning, heating, and artificial light to help ensure reduction of the overall carbon footprint of the city. Meanwhile, the area surrounding the city includes photovoltaic farms, research fields, and plantations, making the community self-sufficient by using renewable energy.

The city is divided into two districts connected by a linear park. The buildings were designed with façades that provide a balance between public spaces and the streets. Some buildings for university students, for example, are clad with terracotta to function as private balconies, while others have metal screens, and some have air-filled wall panels used to decrease the thermal mass of the wall. Public spaces and amenities include a university, public parks, various cafés, organic supermarkets, a bank, and a travel agency.

Masdar City is known to be one of the most successful sustainable urban spaces and the majority of the work should be finished by the year 2030. It is intended to encourage experimentation with emerging technological advances. Expected to house 40,000 residents, this project creates a model for the architects, urban planners, and engineers who are designing tomorrow's sustainable cities.

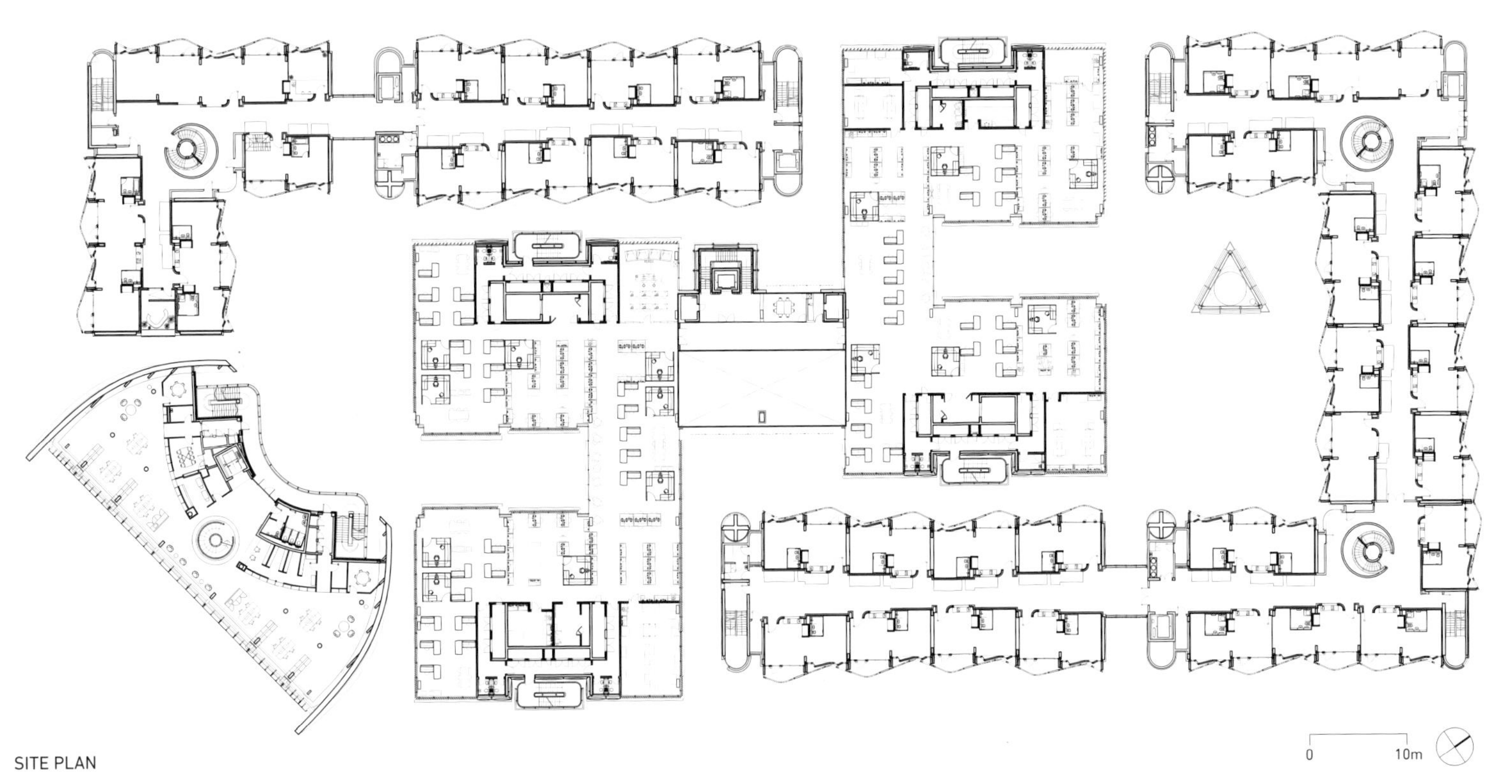

SITE PLAN

大学城

UniverCity

Burnaby, British Columbia, Canada
DIALOG

UniverCity is located on top of Burnaby Mountain in Burnaby, British Columbia, adjacent to Simon Fraser University (SFU) and just above Metro Vancouver. UniverCity offers a wide range of qualities and services that are incorporated into an urban neighborhood surrounded by a forest.

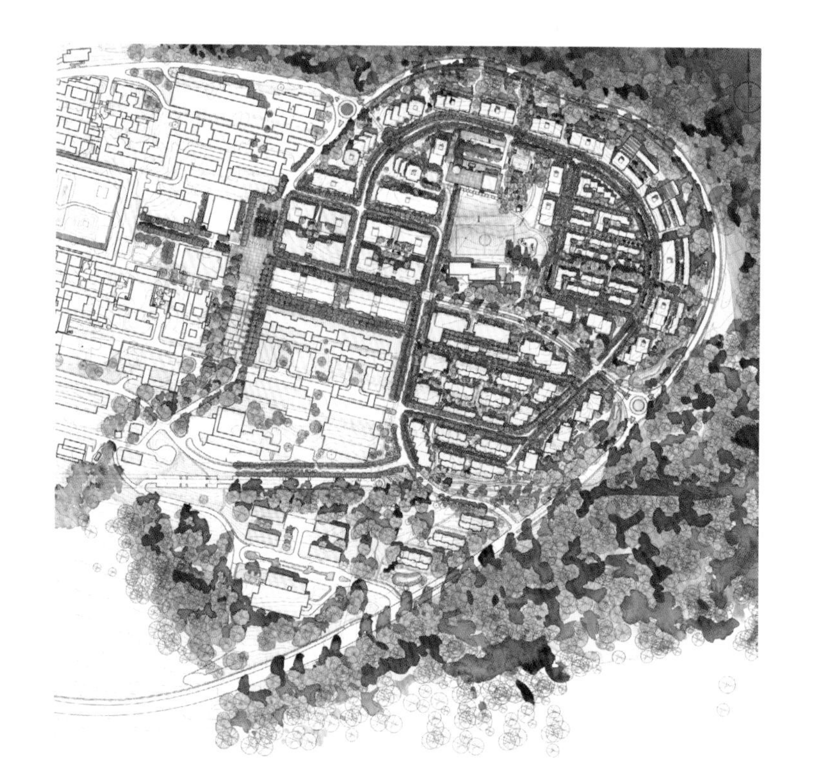

Back in 1963, architects Arthur Erickson and Geoff Massey submitted their initial plans for Simon Fraser University which included a residential community that sat atop Burnaby Mountain. Starting in 1995, Simon Fraser University created the Simon Fraser University Community Trust to guide and oversee the development of UniverCity. In 2000, DIALOG (formerly Hotson Bakker Architects) began the design and master planning of UniverCity with its various zones, transit and circulation systems, and local amenities. With a total site area of 160.6 acres (65 hectares), the master plan is focused on creating a walkable and public-transit-oriented community where residents commute by either walking, cycling, or using the centrally located public transit system.

A detailed watercourse and stormwater management plan, utilizing the first example of bio-mimicry in North America, controls stormwater flows and water quality in order to enhance downstream, fish-bearing watersheds. This unique landscape provides an educational landscape and is monitored by SFU.

The pedestrian and biking network enables the link between residences, retail shops, services on University High Street, University Highlands Elementary School, Childcare Center, nearby parks, and transit bus stops. In addition, car sharing is also an option for getting around: residents can have access to shared vehicles on an hourly or daily basis thanks to the partnership between SFU Community Trust and MODO.

Various amenities that are available at UniverCity include a childcare center, schools, shops, entertainment, and access to lots of green public spaces. Residents are also given access to the nearby Simon Fraser Campus including its art gallery, aquatic center, library, athletics facilities and bookstore. With the help of the "community card," residents can use the libraries, athletic, and

recreational centers free or at a discounted rate. Special events are also offered year round including the Pocket Farmers Market, Summer Busker Series, SFU Theatre performances, and summer camps for young children. Furthermore, a public art program that showcases works by local artists is also offered to residents to enhance the connection between the artist and the individual.

UniverCity offers a wide range of housing types including rental apartments and ownership options totalling 4500 units. Approximately half of the residents are family-oriented, while the others are connected to Simon Fraser University as students, faculty or staff members. UniverCity measures its sustainability with four criteria: "environment" (preserve and improve natural resources on Burnaby Mountain); "equity" (provide a healthy, safe, and affordable place to live and work for all); "education" (enhance student education as future citizens); and "economy" (generate revenue to maximize the long-term value of Simon Fraser University's contributions).

港池7号

Bassin 7 (BSN7)

Aarhus, Denmark

BIG (Bjarke Ingels Group)

BIG (Bjarke Ingels Group) has created a plan for Bassin 7 (BSN7), a new mixed-use neighborhood in Aarhus' harbor front in Denmark. With a total area of 24.71 acres (10 hectares), the importance of this development is the organization of the site's seven residential buildings within a system of recreational and cultural activities, including a beach zone, swimming pools, theater, and café. All are accessible along two different public promenades.

Also known as "Aarhus Island", the project's mission is to create life between buildings where the large open spaces are replaced by intimate street layouts, cozy small spaces, and public squares. This allows for a mix of independent private properties, public life, and social interactions of the bay to take place. The inclusion of waterfront promenades allows residents to reach the boat harbor that is located near the town square and the city center of Aarhus. The first step in planning was to design the public spaces and then carefully combine those spaces with the private residences. The project was meant to "breathe life" into the existing harbor so that the water's edge can be influenced by the public realm. The first of the two promenades will provide a direct route for the harbor-to-city center connection, while the second will criss-cross from residential buildings to amenities such as public pools, a waterfront theater, beach zones, and green spaces. Additional amenities will include cafes, restaurants, kayak ports, and scenic platforms.

Around the promenades, the designers propose to add seven residential buildings that will include over 200 units each. Those buildings are designed to be individual in structure, ranging in height from low- to high-rise. Each building will include a private central courtyard allowing opportunities for privacy within a public realm. The placement of the buildings is meant to minimize noise and create comfortable microclimates. In addition, by including urban activities, the design team has introduced new types of housing to attract families to the area.

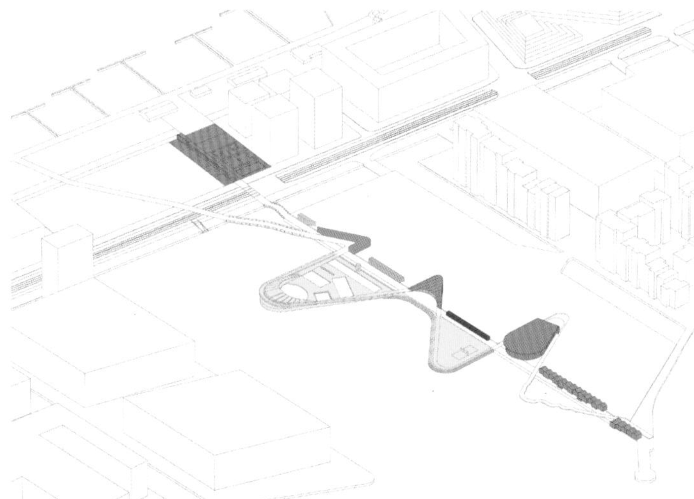

AS AN INSTANT URBAN INTERVENTION, THE PUBLIC PROMENADE IS PROGRAMMED WITH CAFES, RESTAURANTS, A HARBOR BATH, THEATER, AND BEACH HUTS.

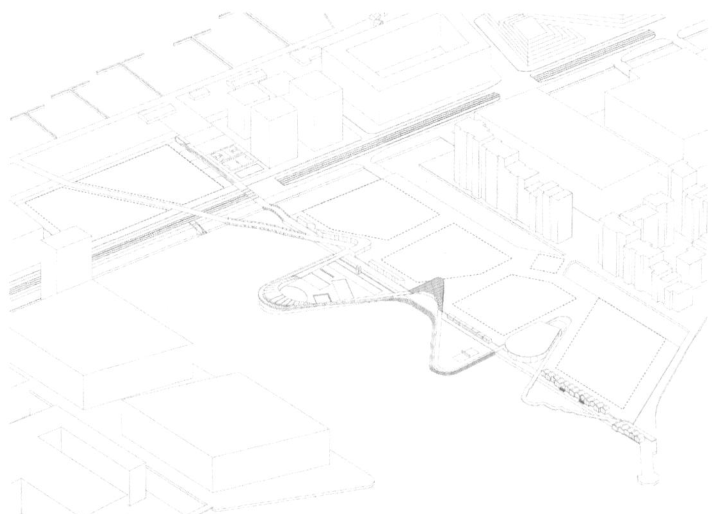

THE LEFTOVER SPACE IS FILLED WITH SEVEN NEW BUILDINGS PLOTS OF DIFFERENT SHAPES AND SIZES.

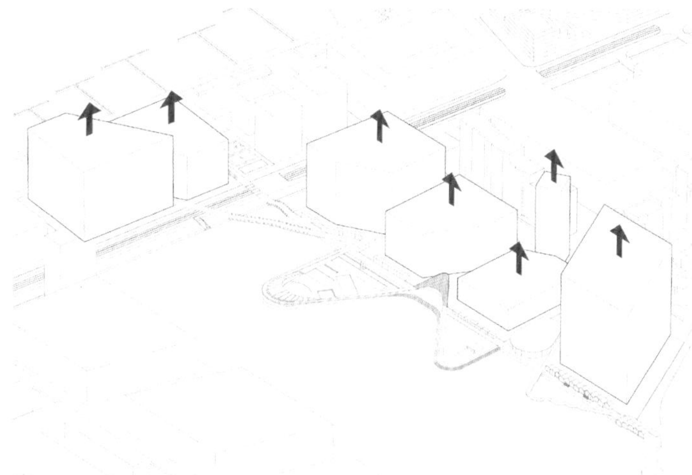

THE BUILDING ENVELOPE VARIES IN SCALE AND HEIGHT, GOING FROM SMALL AND MEDIUM, TO LARGE AND EXTRA LARGE.

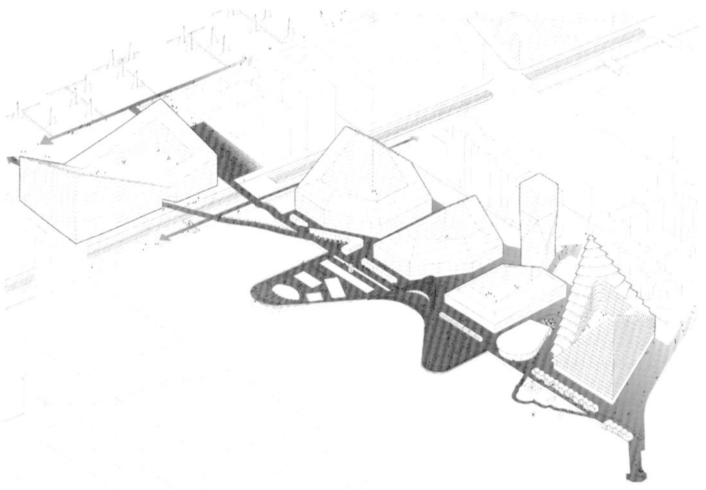

AN INDICATION OF WHERE AUTHENTIC PUBLIC LIFE TAKES PLACE AMONG THE BUILDINGS.

第四章：绿色开放空间

GREEN OPEN SPACES

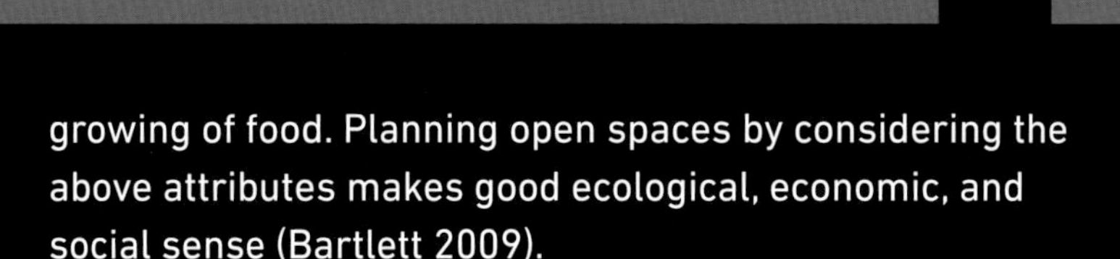

Green open space needs to be planned at the same time as considering existing natural features and the residents' lifestyles. It is likely that a community will have inhabitants of different ages and mobility levels so it is important to plan places that are suitable for children, active adults, and the elderly. Furthermore, a good open space should provide multiple options to suit different moods and activities, weather conditions and seasons. Hard surfaces can be provided for holding a communal yard sale and soft surfaces for casual play, for example. This chapter looks into innovative open-space designs that support active lifestyles as well as the recently designed communities that illustrate them.

In 1733 James Edward and subsequently many others have suggested that trees and parks are the lungs of the city (Piedmont-Palladino and Mennel 2009). Green open spaces beautify a place and in addition offer opportunities for social interactions. They are more than just areas covered with grass and trees. These are the places where children can play and where older people can be active. They also provide opportunities for the growing of food. Planning open spaces by considering the above attributes makes good ecological, economic, and social sense (Bartlett 2009).

Residents commonly value parks, woodland and playgrounds. They appreciate their ability to alleviate stress and the opportunity they offer for exercise, social contact and play (Space 2010). Evidence exists that daily contact with nature such as a simple walk in a natural setting provides therapeutic and restorative benefits. Studies show that residents become mentally and physically healthier the more they are able to live their lives in direct contact with nature. The greener the neighborhoods are, therefore, the more people are motivated to be active outdoors and engaged in social events (Piedmont-Palladino and Mennel 2009).

It was also demonstrated that having green space adds opportunities for exercising for all age groups. For example, at Harbour House in Dorset in the United Kingdom, the front lawn has been turned into an area for playing croquet. The residents combine social engagement with physical exercise when they play (Shackell and Walter 2012). Because of the many benefits, some design firms began to specialize in integrating green spaces in community planning.

CABE Space is an architectural firm located in the United Kingdom that promotes well-designed parks within towns and cities. The firm's motto is that green architecture and design improve the built environment and help create sustainable communities where people want and can afford to live. The firm promotes healthy communities by developing greener and more accessible spatial environments, while ensuring that new developments encourage physical activity and healthy living (Our Areas of Expertise n.d.). Greener communities also have a positive impact on the environment as they reduce the ambient temperature by creating shade for homes during the summer.

It is also important to include green open spaces in communities since they are known to reduce the "heat island effect." The term describes built up areas that are hotter than nearby rural places. The annual mean air temperature of a city with 1 million people or more can be 1.8°F to 5.4°F (1°C to 3°C) warmer than its surroundings. In the evening, the difference can be as high as 22°F (12°C). Heat islands can affect communities by increasing summertime peak energy demand, air conditioning costs, air pollution, and greenhouse gas emissions, heat-related illness and mortality, as well as water quality. Strategically, trees and other plants help to cool down the environment and lower the heat island effect. Shaded surfaces may be 20°F to 45°F (11°C to 25°C) cooler than the peak temperatures of materials exposed to direct sunlight (Bollerud 2013).

Zoning regulations and by-laws should be considered in the design of public green spaces. Innovations such as zoning for open space conversion and drawing boundaries to contain growth allow for more opportunities to positively affect the quality of a community's expansion (Piedmont-Palladino and Mennel 2009). The most appropriate zoning by-laws to be introduced and enforced may depend on the type of residences and environment.

The government amends laws based on the percentage of green open spaces that should be allotted in a community. The purpose of creating a defined green area is to help ameliorate problems of drainage, water quality, and speed of flood waters. Regulations enacted in 2011 by Connecticut in the United States require that a minimum 10 percent of each residential zone should be allocated to green space. This 10 percent ensures a well-maintained community that protects and enhances water and land resources, pervious surfaces, open space, parklands, and recreational facilities in an environmentally sensitive manner (Tesei 2013). While zoning by-laws guaranty green spaces for the community, they do not ensure their maintenance. As a result, further initiatives may be necessary for the proper care of green open spaces.

Once built, green spaces need to be maintained to safeguard their success. Open spaces and natural resources can be protected by preserving and restoring qualities both within and outside urban environments through planned land acquisition programs, such as transferable development rights, protecting agricultural land and forested areas, and establishing high environmental standards (Piedmont-Palladino and Mennel 2009).

There are several basic design principles for green open spaces. They include the location of residences around them to take advantage of the view and the introduction of paths and seating areas according to the site's natural features. Preferably, green space should be located in the center of the community and face the south side of homes, which have more windows, so that residents can enjoy a direct view. Streets should be laid out on the east-west axis to take advantage of the south-facing public space.

When centrally located, the green space becomes easily accessible and within walking or biking distance. Well-designed green spaces need to carefully be planned to preserve and enhance vegetation such as trees or protected species (Shackell and Walter 2012). Creating a detailed analysis and appraisal of the site at the start will yield better results. Paths within the green space should take advantage of the site's natural slopes. They should be far enough apart to allow flat areas where children can play games. This also creates privacy for families wanting to have a picnic. Areas for sitting should be wide and take advantage of views of natural features.

Planning ahead is necessary when creating a communal green space. Walking and biking trails should be considered to create a mobility network. Also, recycling and composting services should be available and easily accessible on-site (Rabinowitz 2013). Furthermore, direct health and safety benefits such as flood control, cleansing of air, and separation from hazards should be part of the considerations as well as measures to prevent crime through environmental design. The designers should look into historic preservation opportunities that remind people of what the place once was (Enger 2005).

Natural open spaces and important features within the site should also be evaluated. The initial site plan should contain analyses of existing and future needs, and retain any natural open space to encourage recreational activity. To further benefit the community, any planning for green open spaces should also consider complementary sets of parks and open spaces that meet the needs of a full range of community interests and work for the majority of people. These large open spaces should be linked to major activity centers such as schools and employment centers by pedestrian paths to encourage physical activity through the green spaces (Enger 2005). A recent innovation and sustainable idea is to incorporate a garden in the green open space for the benefit of all the residents.

In Somerville, Massachusetts, local city initiatives have created the Somerville Community Growing Center (SCGC). A green open space was built for gardening that is run by volunteer garden coordinators who assign plots and help interested residents obtain seeds and compost, as well as providing information and advice. The SCGC is located on a quarter-acre parcel and has become much more than a community garden, offering both educational and cultural programs. Solar panels located throughout the garden provide energy for lighting the barn and pumping the fountain and stream. Water used for washing hands is cleaned and recycled to water the garden. All gardening is organic and complementary plantings are used to encourage healthy growth. The range of produce varies from year to year, but the fruit sampler lists peaches, apricots, grapes, sour cherries, elderberries, blackberries, raspberries, blueberries, and sometimes strawberries. Vegetables varieties include tomatoes, lettuce, and cucumbers, garlic, hearty greens, and hot peppers (Model Right to Farm By-Law n.d.). The opportunity for community farming is made possible by the Right to Farm by-law.

The Right to Farm by-law encourages the pursuit of agriculture, promotes agriculture-based economic opportunities, and protects farmlands by allowing agricultural uses and related activities to function with minimal conflict with abutters and local agencies. This by-law is an important tool that can bolster a community's efforts to protect the viability of farming. The intent of such a legislation is to reiterate the importance of, and support for, farming within the town. There is a notification provision that works to make sure that people moving into the community are aware that agriculture, and its associated sights, sounds, and smells, is an accepted and central economic and cultural activity. This type of by-law seeks to prevent conflicts between gardeners and neighbors by establishing a dispute resolution process (Model Right-to-Farm By-Law n.d.).

In the planning process of a community, careful attention should be paid to green open spaces as they ensure a healthier lifestyle and a lower carbon footprint. The residents also appreciate green open spaces since they not only improve their physical and mental health but also foster social engagement.

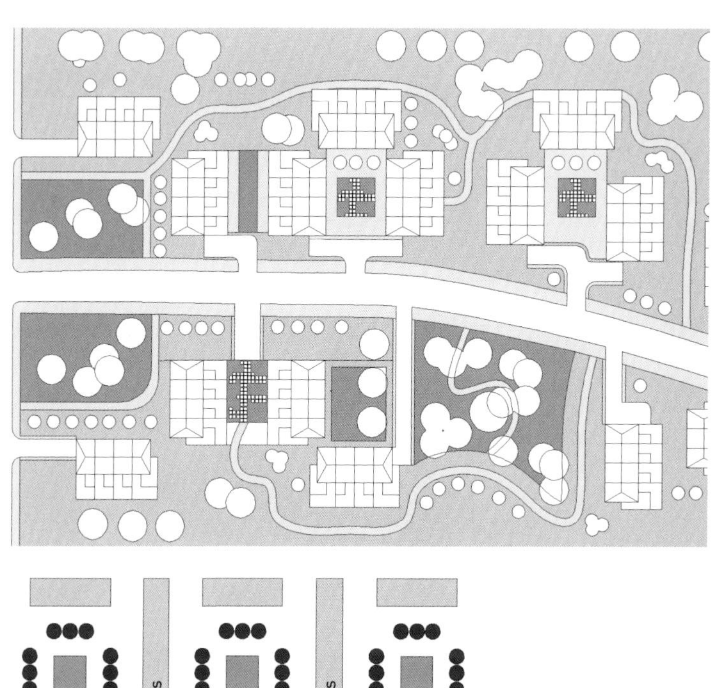

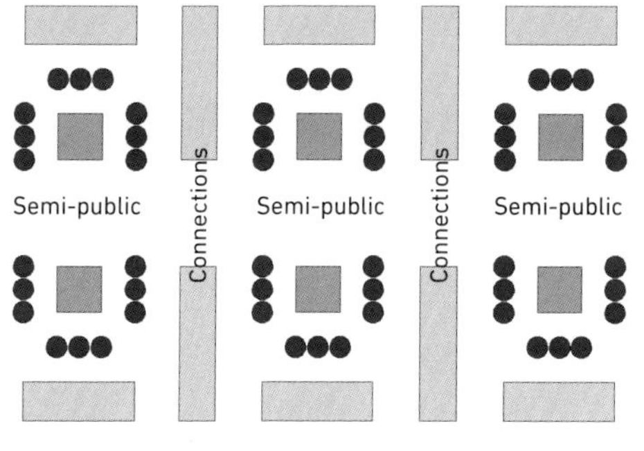

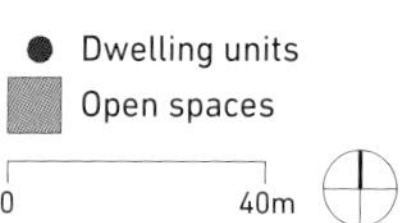

A PUBLIC OPEN AREA NEEDS TO OFFER A MIX OF LARGE AND SMALL SPACES TO ACCOMMODATE FOR A RANGE OF ACTIVITIES.

海勒街公园和住宅

Heller Street Park and Residences

Melbourne, Victoria, Australia

Six Degrees Architects

The Heller Street project is an excellent example of how green space can take precedence in community planning. The city council of Brunswick, a suburb of Melbourne, had a surplus of land. Despite a public wish to use the land as a park, the council could not take on the financial risk of developing it with public money. To appease both residents, and its own financial agenda, the council approved private development on the land, but stated that only one-third of it could be used for residences, the other two-thirds must serve as a public park. This allowed room for 10 residences, built with young families in mind. The agreement also stipulated that the parkland must be decontaminated, as it was formerly used as a public waste yard.

The ambiguous nature of public versus private space is what distinguishes this project from others. The park serves as a beautiful front yard for residents, but it is also open to the public. Placing the public space directly on the street, rather than in between residences provides a service to the community as a whole and makes the project feel less exclusive. Had it been placed between the residences, the general public would feel unwelcome in the development. Since the land is maintained by the council, it offers the residents hassle-free access to green space. Roofs collect stormwater runoff, which is used to water the space. This guaranties that even during a drought, the park will be pristine and healthy.

This arrangement benefits both the council and the property owners of Heller Street. The council cuts costs by allowing a private company to develop the land, and the residents have access to better amenities. Some efforts have been made to create a distinction between private and public land using the natural landscape rather than harsh barriers. The park is surrounded by a circular 4-foot-high (1-meter-high) mound. The mound was created using the soil excavated when building the residences' semi-basements. The landscape architect, Simon Taylor, intended the mound to be an elevated surface where the public could sit and relax. The mounding also provides mediation between the residents and the public park. The vegetation, now established, helps with this visual separation.

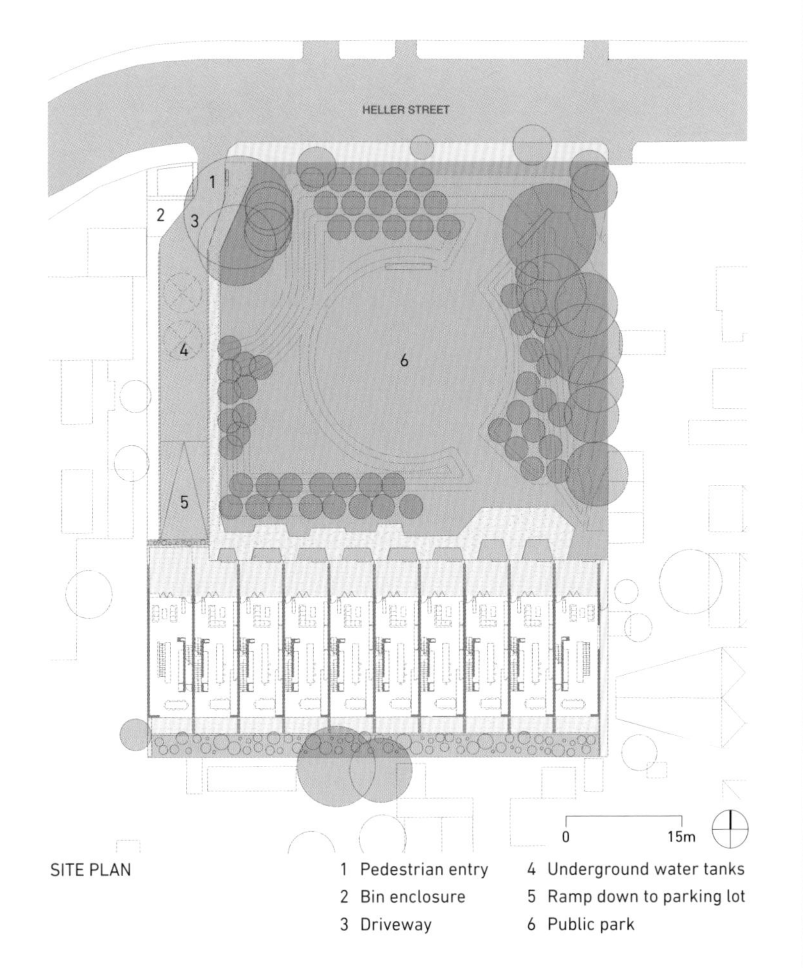

SITE PLAN

1 Pedestrian entry
2 Bin enclosure
3 Driveway
4 Underground water tanks
5 Ramp down to parking lot
6 Public park

As an alternative to a private yard, residents can take part in community-wide outdoor activities such as gardening, kicking a ball, and even pop-up moonlight cinema screenings. The shared open space in front of their homes and participating in the shared activities offer valuable alternatives to outdoor land ownership. This creates an improved sense of place and allows citizens to contribute to a larger purpose. This masterful blend of public and private space saves resources, allows for higher unit density, and creates more accessible green spaces, both for residents and those in the neighboring community.

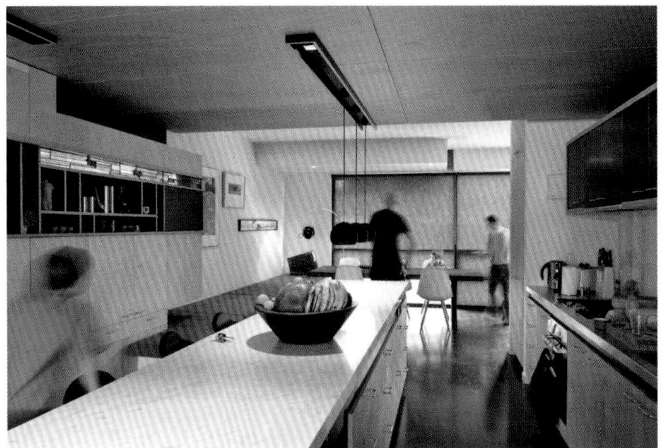

温嫩登乐土社区

Arkadien Winnenden

Stuttgart, Germany

Ramboll Studio Dreiseitl (formerly Atelier Dreiseitl)

Arkadien Winnenden is an industrial renewal project in Winnenden, a suburb of Stuttgart. The economically depressed town was once a large industrial center, but due to the closure of many factories the future of the area seemed bleak. The new development was meant to provide the town's residents with a brighter future, placing sustainability and community values at the project's heart.

The former brownfield site is within walking distance of many amenities. Pedestrians can reach the town center and train station within 5 minutes, and there are nine schools within a 10-minute walk. Downtown Stuttgart can be reached within 30 minutes by train.

A dense layout allows for a smaller site and creates pleasant, village-like streets. It has also allowed the developer to create more housing, at an affordable rate. Recycled granite, as well as cost-efficient concrete pavers and asphalt, were used to create the streetscapes. The percentage of impermeable surfaces in the area was cut down from 95 to 30 percent, using permeable pavers and structural substrate. The substrate creates garden-like parking spaces, blending the parking lot with the natural landscape.

The pedestrian-friendly design of the community places social sustainability at its forefront. Street corners serve as places of interaction and have been designed as public centers, rather than mere places of transit. Streets have limited access for cars, and parking is provided underground or in designated spots around the city. This creates a safe environment where children can play, with cars hidden from view so that they do not detract from the natural landscape. The barrier between public and private realms has been made ambiguous, and therefore fosters a healthy communal living environment.

Creating the perception of a less-dense neighborhood is made through landscaping and gardening techniques. Gardening is focused on native plants, which provide a healthy ecosystem for birds and bees. Green roofing has also been used to give prominence to the natural landscape of the community.

A beautiful lake has been placed at the center of the development to act as the heart of the community and offers residents a place to relax and socialize. The lake also serves as a rainwater detention basin, filtering the water using a stepped system. The water then flows into a flood meadow and enters an environmentally renovated stream. A flood retention system has also been put into place, in case of storm flooding.

Arkadien Winnenden is an example of sustainability and economic regeneration working hand-in-hand. The developer has taken measures not only to secure themselves financially, but also to give back to the residents of a demoralized town, and create a sustainable community.

布里克近邻社区

Brick Neighborhood

Ljubljana, Slovenia

dekleva gregorič architects

The mandate of the Slovenian-based firm of dekleva gregorič architects was to create a neighborhood with a clear spatial, material, and social identity. This needed to be addressed through a unique design and spatial organization of the dwellings and their surroundings in order to establish a connection among its present and future residents. The creation of a diverse, public open space, which caters to all individuals, is crucial when meeting basic community development aspects.

With an area of 5 acres (2 hectares), a public open space was planned in between the apartment buildings. Therefore, communication between the two sides that make up the community is strengthened and further motivated by this space. Benches and a children's playground were introduced as well. One unique aspect of this communal space is the inclusion of poems by Slovenian poets that feature on objects such as playground equipment, benches, and doors. Each communal space is culturally enriched by the chosen poetry, establishing a unique identity and increasing cultural awareness among the residents.

The spatial organization of each of the 185 apartment units creates 17 different types of dwellings, ranging in size, volume, internal arrangement, and circulation—all of which can meet the needs of present and future inhabitants. Furthermore, each unit also has a unique play area with both external and internal aspects. Articulating the external brick façade to give a distinct look, for example, gives each apartment its own identity and allows for further community distinction. The use of brick is a historic tribute to the site's former function as an industrial brickyard. Corridors among the apartment units are designed as places of meeting between the two sides. Common spaces above the buildings' entrances also accommodate social interaction. These spaces illuminate the possibility for children's birthday parties, indoor playgrounds for the winter season, and exercise classes.

The design demonstrates that highlighting the visual identity of each space by using different materials and unique artistic expressions can contribute to social cohesion among residents.

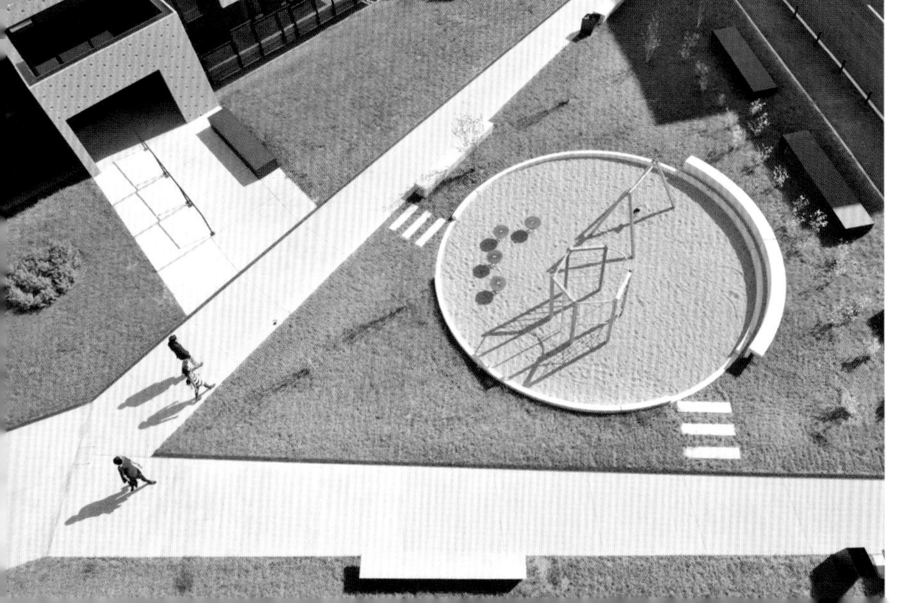

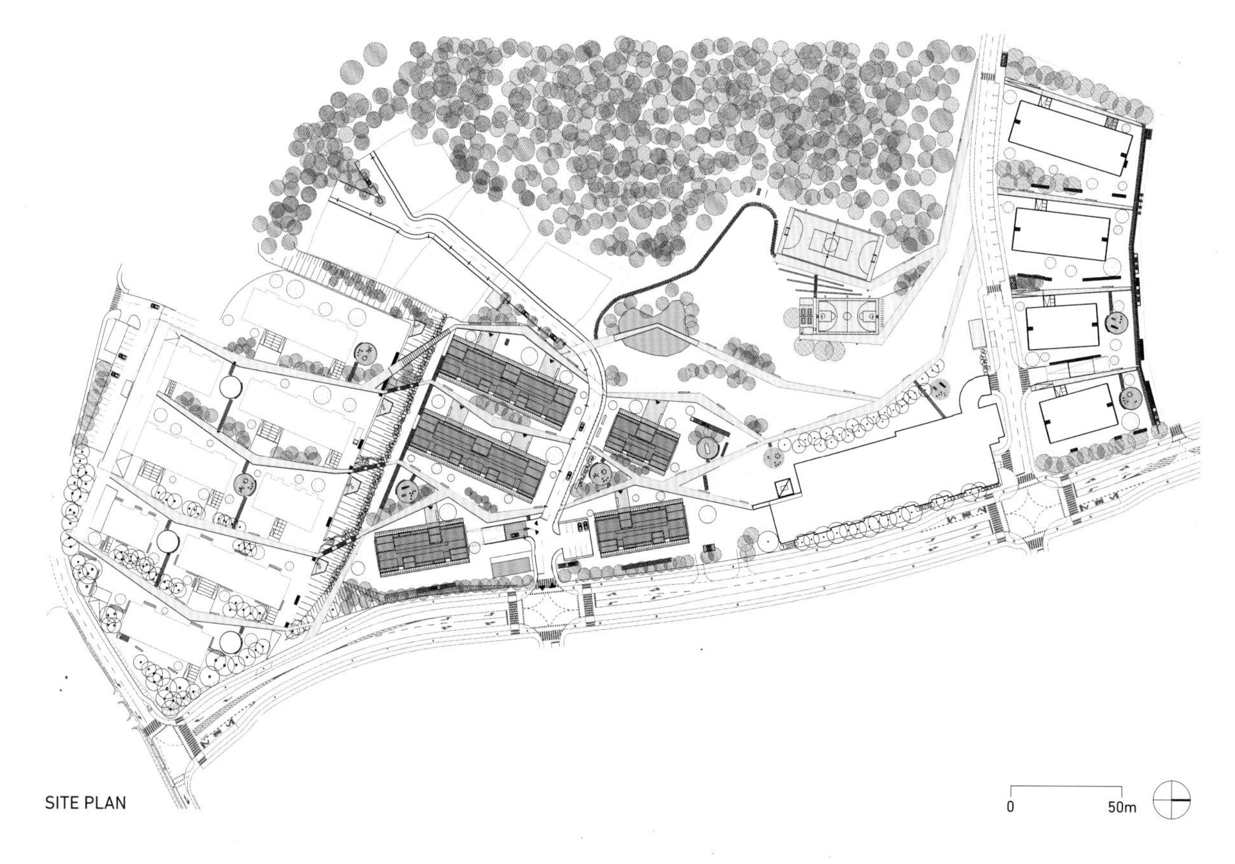

SITE PLAN

阿尔博莱拉生活社区

Arbolera de Vida

Albuquerque, New Mexico

Design Workshop

Arbolera de Vida was one of the first Spanish communities in the historic city of Albuquerque, New Mexico. In the 1980s one-third of its population lived below the poverty level and experienced many hardships such as crime and poor air quality, due to a local sawmill. After a lot of pressure to close the mill, the landscape architectural firm, Design Workshop, was awarded a mandate to re-build the community so it would be financially stable and to limit the gentrification of the community by using sustainable design strategies so as to keep housing and operating costs as low as possible. Currently, this newly improved development is recognized as a pleasing, and more importantly, an affordable place to live.

Design Workshop worked with local residents, merchants, and the city of Albuquerque to come up with an innovative design scheme. Design Workshop's master plan placed particular focus on integrating social, ecological, and economic aspects. This design comprised a total area of 27 acres (10.92 hectares), located just south of the I-40 and north of historic Albuquerque, and was based on residential, commercial, and industrial zoning. Working alongside the Sawmill Community Land Trust (SCLT), priority was given to creating affordable housing and the total restructuring of the economic infrastructure, providing low- to moderate-income families with a comfortable place to live. With two out of three construction phases complete, a total of 23 owner-occupied homes, 67 affordable homes, and 60 affordable living-and-working units of various types such as single-family detached, duplexes, and townhouses were built.

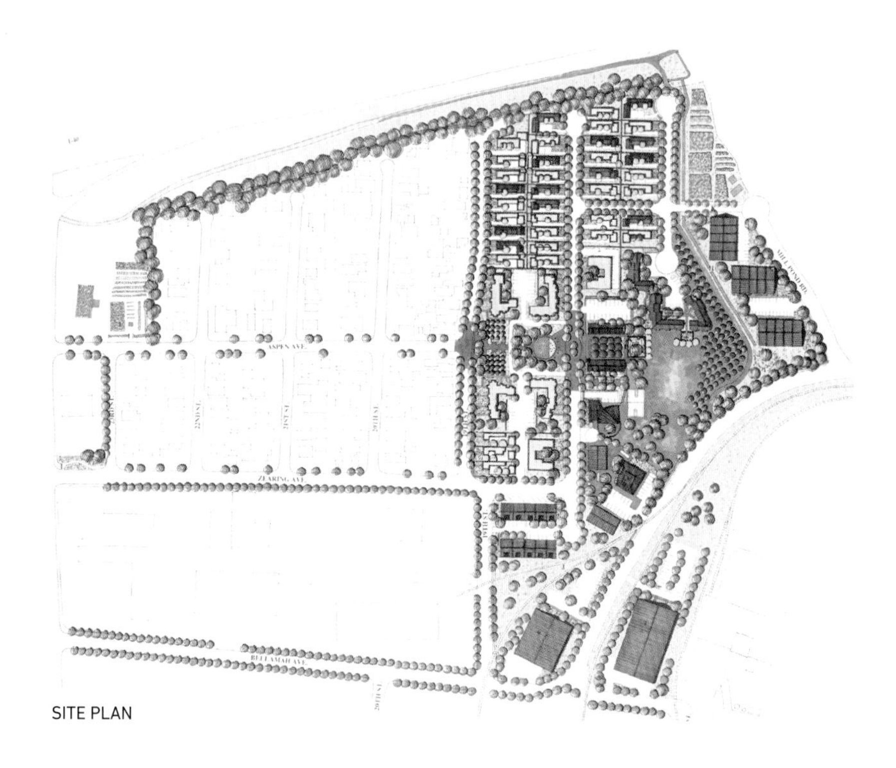

SITE PLAN

ARBOLERA DE VIDA

The mission of the re-design of Arbolera de Vida is to re-establish and motivate the once lost social connections by providing plazas, parks, playgrounds, and community gardens. The multi-purpose plaza, encompassing a total area of 1 acre (0.4 hectares), was the central point for the first phase of construction. The design employs vivid Mexican colors, curvilinear shapes, multiple levels, textures, and architectural aspects in the design to bring out the messages of "community" and "invitation." A 2-acre (0.8-hectare) park was also built in the second phase and includes walking and bike pathways.

In the third construction phase there are more housing units, shops, and a community and senior center. Furthermore, a working "acequia" (Spanish for irrigation channel) will be included in a community garden in the near future. Nearing the end of completion and seeking increased community collaboration, this re-defined neighborhood is continuing to grow and earn substantial recognition for bringing success to what was once an inhospitable place to live.

AIRPORT PARKING
per day EVERY DAY!

协和社区

Accordia

Cambridge, United Kingdom

Feilden Clegg Bradley Studios, Grant Associates

Urban designers and architects at Feilden Clegg Bradley Studios along with landscape architects at Grant Associates were given the task to create a different and more substantial style of living in Cambridge, United Kingdom. The mission of this project was to create a pleasing urban place that balances usable private space and a high-quality public setting. Various types of housing, including apartment units and large-scale housing integrated in public areas, reflect some of the main features of this scheme.

The overall design context of Accordia is the natural setting containing over 700 mature trees encompassing an area of 8.6 acres (3.5 hectares) of the total site of 23.7 acres (9.6 hectares). These trees are an essential part of the communal garden spaces in the creation of this high-density housing complex. Public landscape spaces are presented as potential places for various functions and uses based upon different residents' preferences. For example, grand open spaces provide potential for play and leisure, while the garden spaces propose different themes such as gardens outside the kitchen as places for long walks, or part of the center lawn.

Re-designing the pedestrian paths, bicycle routes, areas for bike and car parking are also nestled in the landscape as shared public spaces and link all the landscape areas in a single network. All of the paths are color or texture coded to coincide with the architectural materials of the overall design. The design of the housing blends the traditional domestic style of Cambridge housing with modern elements in a unique landscape setting.

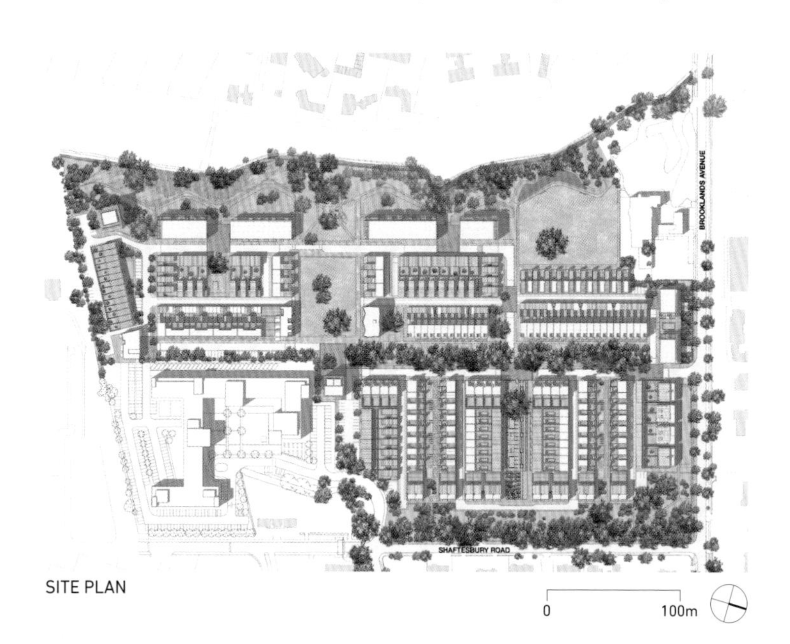

SITE PLAN

Traditional gardens were arranged in various forms in private open spaces such as courtyards, roof terraces and large balconies. Terraces and courtyard houses are formed by weaving houses and apartment units.

A total of 212 houses and 166 apartment units with a density of 40 dwellings per hectare were strategically organized in Accordia. The unique identities of the dwellings were based upon their layouts, which range from simple to highly complicated two-level units that require the integration of complex staircases. Their individuality was also shaped by the use of various external materials, ranging from brick similar to the traditional Cambridge Gault clay bricks for the houses, and copper and green oak for the apartments.

Feilden Clegg Bradley Studios and Grant Associates design for Accordia was highly praised and is considered as the front runner for new large-scale housing design and planning in the United Kingdom. It was the first housing scheme to be awarded the RIBA Stirling Prize in 2008. It has been stated that more British communities need to follow Accordia's example—where a faintly controlled plan enables the blending of architecture with the landscape.

第五章：密集型住宅小区
COMPACT NEIGHBORHOODS

In recent years, it has become abundantly clear to policy makers and planners that the practice of urban sprawl needs to be halted and the building of denser, more compact communities must be considered. In addition to the environmental degradation caused by urban sprawl, the demographic transformation is another argument for introducing neighborhoods with higher densities. This chapter describes such places and suggests guidelines for the planning of livable compact neighborhoods.

Compact neighborhoods have a high number of dwelling units per area (Kackar 2003). The indexes of density are gross density that is calculated by dividing the total number of units by the site's total area. Net density is calculated by dividing the total number of units by the total area excluding the roads and open spaces. A typical North American suburb commonly has four to seven units per 1 acre (0.4 hectare), whereas in high-density neighborhoods, there is on average a minimum of 35 units per 1 acre (0.4 hectare).

In general, urban density is a subjective term that relates to a particular location and culture. An Asian neighborhood is likely to be much denser than its North American counterpart. The question is, therefore, how should one regard density and what are the common yardsticks for such neighborhoods?

Typical 20th-century suburban and city forms have two distinct densities. The first is considerably lower in density, and averages 7 units per acre (17 units per hectare). The large lot size, land, and infrastructure costs per unit are high resulting in higher dwelling costs and, effectively, fostering suburban sprawl. In contrast, the higher density of a city averages 31 units per acre (77.5 units per hectare), which is likely to be unpopular with the typical buying public. Would-be buyers will be reluctant to accept crowded communities which lack public and private open spaces for example. By combining planning features from low-density and high-density designs one can introduce urban forms that mix both. This design averages 22 units per acre (55 units per hectare), with rear private parking and yards for each unit. Minimal, though acceptable, widths separate the houses. Moreover, green open space located at the center of the cluster is accessible from each unit and is associated with the view that public parks are crucial to community interaction. The new design for high-density communities mixes ideas taken from traditional, late 19th-century high-density communities with contemporary designs.

Compact neighborhoods have several advantages. First, the number of people who reside in them can support local commerce. Research suggests that communities with a minimum density of seven units per 1 acre (0.4 hectare) or higher are needed to support a corner store and communities with eighteen units per 1 acre (0.4 hectare) are required to support a small supermarket. On the other hand, in typical neighborhoods with densities of four to seven units per 1 acre (0.4 hectare), residents will need to drive to basic amenities. Compact housing reduces the driving distances and increases walking and biking opportunities. In addition, with fewer roads more land will be available for green open spaces.

It was also demonstrated that residents of high-density neighborhoods have the lowest tax rates since the maintenance is shared among many people (Kackar 2003). Moreover, for municipalities a road with water, sewer, gas, electrical, and phone lines provides four times the return on investment if it services 32 units per 1.0 acre (0.4 hectare) versus eight units per 1 acre (0.4 hectare) (Piedmont-Palladino and Mennel 2009).

Recent social changes have brought about renewed enthusiasm from architects, town planners, and housing officials for compact neighborhoods. The small footprint that this style of community occupies allows a sustainable approach to design and energy-efficient, affordable living. With the changing demographics of buyers in the market and the increase in the number of non-traditional households, the design of these communities has gained the interest of land developers as well.

Five principles guide the design of successful compact communities. First, the density should be increased proportionally according to location. Second, people and places should be connected through a complete street network that promotes walking and biking while including public transit. Third, amenities should be mixed within the community so that the residents can choose to live near jobs, libraries, or schools. Next, public parking is strongly recommended for high-density neighborhoods and should be located in different areas rather than in a single place to create appealing communities. Finally, the planner should include interesting public gathering places.

As noted above, in high-density projects careful attention should be given to roads and parking because they affect the overall cost and contribute to the curb appeal of the community. It is often argued that high-density planning needs to begin with a consideration of parking alternatives. To begin, parking areas should be located at the rear of clustered houses to avoid negative visual effects. Since in most places by-laws require allocation of parking spots for visitors in addition to residents, visitors can use existing roads to reduce the area of asphalt needed, and therefore reduce the urban heat island effect (Bollerud 2013). The planner should also integrate and prioritize pedestrian and biking paths. Pedestrian paths should be well-marked, well-lit, and elevated from the parking level (Kackar 2003).

High-density neighborhoods are beneficial to the environment because clustered dwellings preserve natural areas. When a smaller area is used for buildings, there is less disturbance to fauna and flora. Joint units with a common wall also reduce the materials used in construction. The relationship of these houses to each other is also significant when it comes to energy management. For example, a semi-detached structure with two narrow houses is 36 percent more energy efficient than a detached house, while a unit in a row can be up to 64 percent more efficient than a detached house (Friedman 2005). This energy reduction can also be applied to air-conditioning, which accounts for a large portion of the energy consumed.

Another aspect that occupies both housing providers and consumers is affordability. It has become very difficult for first-time homebuyers to purchase a dwelling in most urban centers. Due to their physical characteristics, these narrow-front homes reduce the amount of land consumed as well as the investment in costly infrastructure which makes them affordable.

The result of compact neighborhood planning should be a mix of uses to serve the residents, a range of housing types and prices to support the diversity of the community, and green open spaces to provide healthier living environments. Linderoth Associates, an architectural firm from Arizona in the United States, introduced alternative design solutions that offer a range of housing types, densities, pricing levels, and lifestyles for sustainable and diverse communities. The firm creates walkable high-density multi-family housing that includes stacked flats, townhouses, green open spaces, and well-integrated amenities (Linderoth 2010). The Aggie Village development in Davis, California, which was designed by the firm, is typical of many higher density projects that provide a variety of housing types, including single-family, duplexes, and accessory units (Kackar 2003).

In Portland, Oregon in the United States, the density levels of the neighborhoods vary from high to low. The driving distances and traveling methods were compared in a study for the different districts of the community. Marked differences were found. Specifically, people in the denser section were about 11 times more likely to use public transit and about 11 times more likely to walk, as well as about twice as likely to ride bicycles. It was found that on average, individuals in denser mixed-use communities drive about half as many miles and have half the level of car ownership compared to the average in the rest of the region (Piedmont-Palladino and Mennel 2009).

Aaron Cheng's study titled *A New Neighborhood for Future China* (2007) looked at the idea of a high-density neighborhood that includes low-density housing for more diversity and better accommodation of different families. In China, building codes allow neighborhoods to reach a certain FAR (floor area ratio) and height limitation, creating conformity and leaving architects with fewer design choices. The buildings have no identity and the communities have lost their diversity. To solve this problem, high-density and low-density housing should be integrated in the community to create architectural diversity.

Cheng suggested that the original, typical neighborhood be replaced by a mix of densities, which provides the opportunity to create diverse private and public spaces (Cheng 2007).

Another example is in Mountain View, California, which was completed in 2000 by the firm Calthorpe Associates. It is a compact housing project with a net residential density of 30 units per 1 acre (0.4 hectare) and gross density of 21 units per 1 acre (0.4 hectare). It has, in total, 102 single family detached houses, 129 row houses, 128 condominiums, and a public parking lot containing 200 spaces for roughly 1000 residents. The housing types range from a density of 11 units per 1 acre (0.4 hectare) to 70 units per 1 acre (0.4 hectare). The part of the community with the lowest density is located furthest form the central amenities, but still within walking distance.

Most residents live within five minutes of the Caltrain rail station, community pool, public parks, retail, offices, and other mixed commercial buildings. All the parking for the units is either located behind, deeply set back from the housing fronts, or underground. The main streets are made narrower to make it possible for visitors to park in the streets. The streets are lined with trees to provide shade for pedestrians. To further protect the environment, existing redwood trees were preserved during construction (Kackar 2003). The outcome is a walkable neighborhood connected by commercial and residential buildings through new public transit. It offers diverse, comfortable, and livable housing at moderate prices in a very dense manner, transforming the community into a sustainable environment.

High-density communities are becoming more and more common. In order to mix housing and amenities, integrate green open space, and provide public transportation within a walkable radius of most residences, particular care should be taken during the design process of compact communities. If achieved, these factors ensure healthier lifestyles, guaranty a lower carbon footprint, and promise a prosperous, sustainable environment for residents.

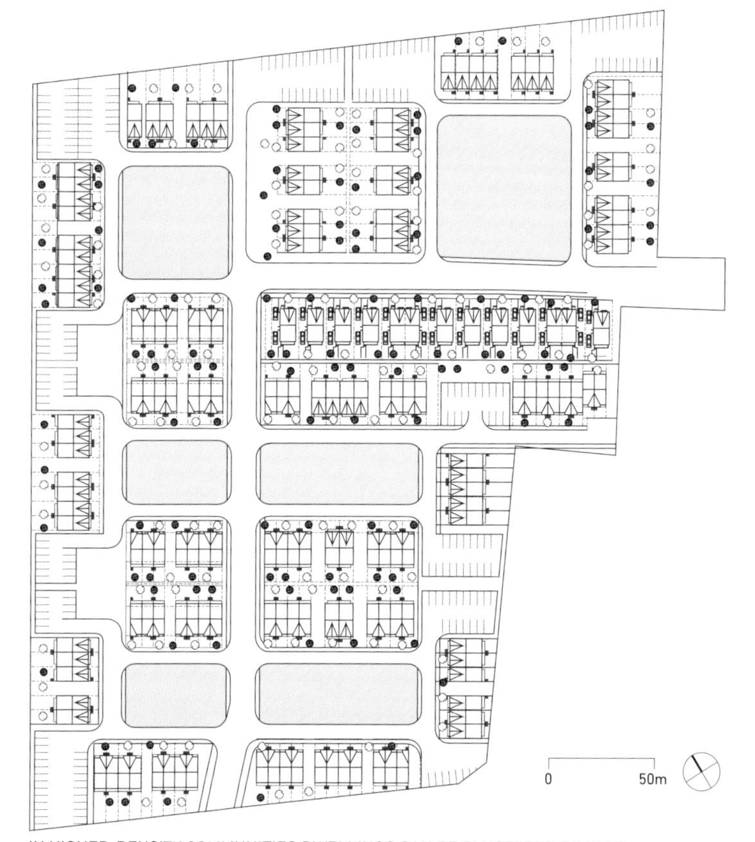

IN HIGHER-DENSITY COMMUNITIES DWELLINGS CAN BE CLUSTERED TO FACE GREEN OPEN SPACES IN ORDER TO ALLEVIATE A SENSE OF OVERCROWDING.

克里斯蒂·沃克社区

Christie Walk

Adelaide, South Australia, Australia

Ecopolis (Dr. Paul Downton)

Christie Walk is a unique housing development within walking distance of Adelaide's city center in Australia. Based on Paul Downton's *Ecopolis* vision, which promotes ecologically designed cities, buildings, and landscapes, the 1990s Christie Walk project was inspired by the need to create ecological cities; the development of agricultural sustainable ecosystems for self-sufficiency. Downtown Adelaide recognized that with the increase in population comes an increase in the need to develop accommodating cities, therefore ecological restoration needs to take place.

The design of the 0.5 acres (0.20 hectares) Christie Walk is an excellent example of compact planning. With such a reduced area, the overall footprint of the development needed to be small. Defined as an inner-city site, its proximity to the center of Adelaide enhances the motivation for residents to walk rather than drive. Public access and curiosity about Christie Walk is welcomed due to the communal pathways located at the front of the townhouses. Due to budget constraints and reducing construction costs, Downton asked local residents to help with constructing the village landscape where curved pathways and paving circles were created from re-used site materials. Residents also helped plant garden beds, which included native plants with low water requirements and exotic trees located just outside the housing units for shade. To further decrease the environmental footprint of the village, the residents were encouraged to grow food on a communal raised garden bed. Stormwater is also retained on location in water tanks just below the automobile parking areas.

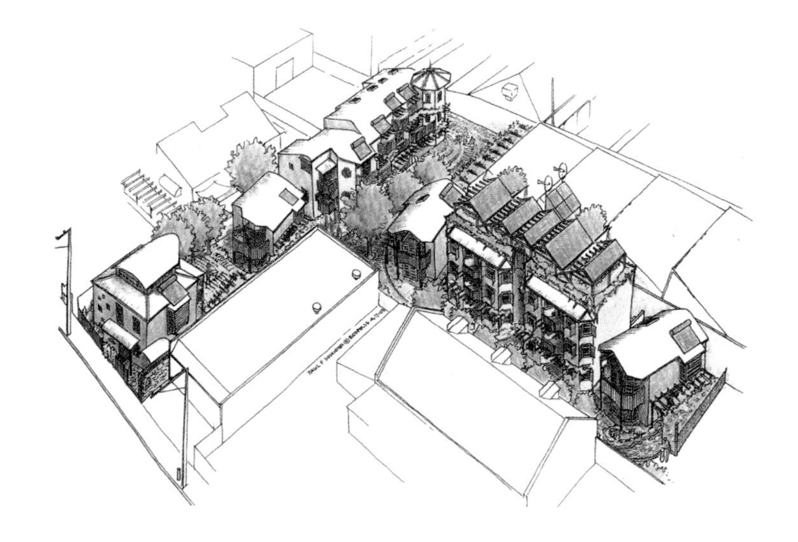

With a total of 27 housing units, Christie Walk includes various types of housing including a three-story block of a six apartments, five townhouses, and three cottages. In keeping with the demands of a growing family, individual dwellings contain two or three bedrooms. In organizing the spatial arrangement of the units, solar orientation was enhanced as much as possible. Building materials such as aerated concrete blocks, straw bales, structural timber, and marmoleum were used for their non-toxic properties and because they help reduce thermal mass. Passive-energy techniques, such as the installation of windows and vents to allow for passive cooling, were also put in place. Furthermore, solar panels were fixed to the roofs to decrease the reliance on exhausting active energy sources. In showcasing Paul Downton's vision of sustainable design, private roof gardens were constructed atop the apartment blocks. These created extra outdoor space in an already dense urban fabric, offering panoramic views across the dense village and adding another layer of insulation to the unit. The design limited the total parking lot area to 11 spaces at the site's edge, thereby making room for more greenery.

Paul Downton's design for Christie Walk achieved wide recognition and won an international environment award for good practice in 2006 when the two first phases of construction were completed. The success of this development comes from the lifestyle choices of the architect who demonstrated that it is possible to construct a lively inner-city community in a limited space.

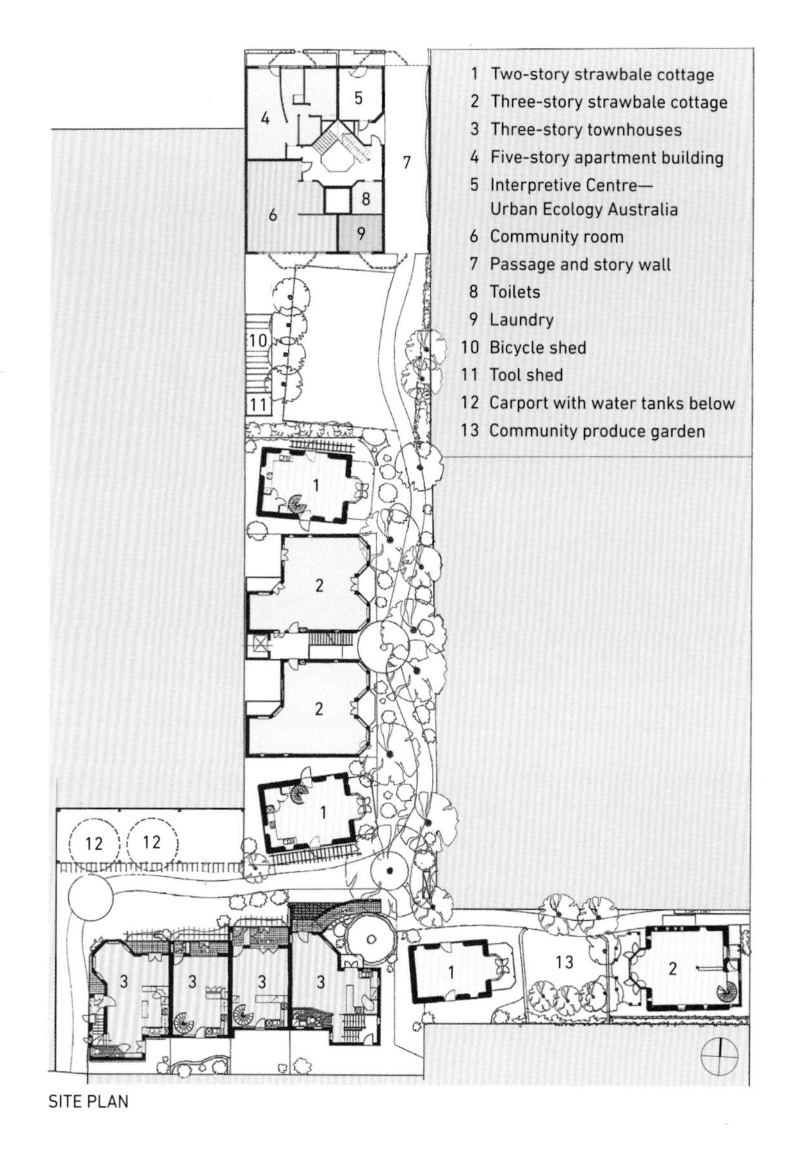

SITE PLAN

20

丐军庭院社区

Geuzentuinen

Amsterdam, The Netherlands

FARO architecten

The suburbs of western Amsterdam are currently in the process of a re-design due to a housing shortage. The post-war western garden suburb of Geuzentuinen is just one of many communities undergoing this process in order to meet present housing standards. The design by FARO architecten features three blocks with a total area of 2 acres (0.8 hectares) that are strategically centered on public, communal courtyards, and organized to allow optimal sun exposure for all buildings. A total of 138 units that vary from a tower to low-rise apartment dwellings occupy this small area.

The main goal for this redevelopment was to establish a modern garden community that was sustainable and energy efficient. To reach this goal, certain objectives had to be followed such as: augmenting the present garden city guidelines, defining green spaces among the community, developing communal gardens, providing constructive solutions for parking issues, allowing for architectural and spatial expression in the design, and mixing rental and owner-occupied housing units at a high density. To achieve such density, the architects divided the site into three groups to replicate the existing patterns of the surrounding streets. Each group includes a combination of rented units that are either stacked or ground-based, and owner-occupied units, which are either taller or shorter volumes—all arranged around a public garden. The stacked apartment dwellings are oriented for maximum sun exposure in areas with less obstruction, whereas the tallest buildings are facing north. Beneath the inner courtyards that include a public garden, a partly underground parking zone is provided for each group of three units. A wall positioned around the garden divides this private shared area from the external surroundings beyond the community.

The façade's materials consist of aluminum on the inner façade with brick covering the outer. All units have a flexible floor plan and street entrance so that different families can adapt spaces for their specific needs. The largest number of units features an entrance with no hallway in order to preserve as much space as possible, and so that inhabitants and guests are at the core of the house upon arrival. The units that include their own parking zone have a split-level plan, which gives the residents 13.1-foot-high (4-meter-high) upper stories on the street side of the site.

The redevelopment of the Geuzentuinen community created a link between the existing development and the modernist vision of post-war garden communities to meet contemporary housing standards in a sustainable way.

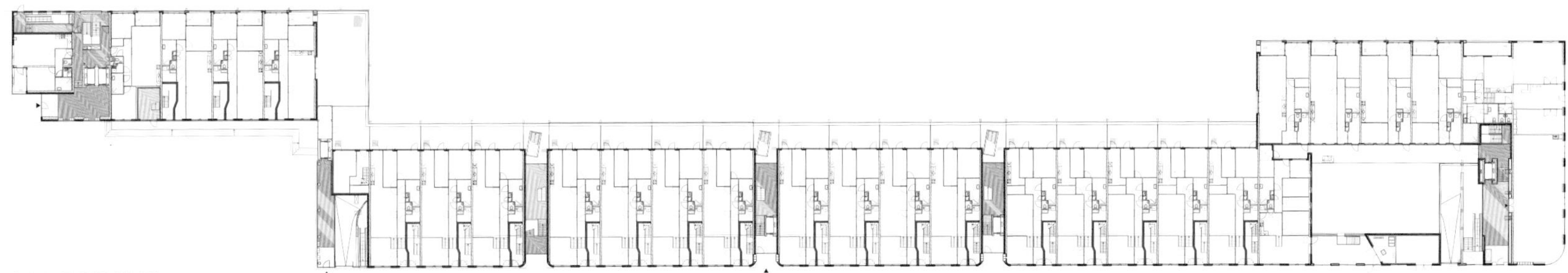

GROUND-FLOOR PLAN

帕克兰公寓

Parkrand

Geuzenveld, Amsterdam, The Netherlands
MVRDV

The re-design of Buurt 9 (Neighborhood 9) forms a large part of the post-war garden cities west of Amsterdam. The new development called Parkrand features the re-organization of the housing units into one compressed but functional volume. By doing this, more area for the surrounding green public park and communal spaces is preserved.

The previously existing Buurt 9 featured 174 small dwellings, arranged in three L-shaped complexes that were located in proximity to a small public park. MVRDV's Parkrand tackles the limitation of small space and high density by making the 200 housing units fit in one volume so that the nearby park and communal areas become more spacious. With a total site area of 3.5 hectares (8.64 acres), the block-shaped complex's dimensions are 443 feet (135 meters) in length, by 111.55 feet (34 meters) in height, and 111.55 feet (34 meters) in depth.

The building program of the complex consists of five towers occupied by a total of 240 residential units that are connected by a shared public patio and a sequence of rooftop penthouses.

The reason for implementing this connecting space is to create the illusion of a hollow block with a panoramic view. Semi-public balconies located above the public patio, on the other hand, allow privacy and views of the nearby park.

The raised public patio between the residential towers permits spaces for gardens, safe entrance, social spaces, and even playgrounds for children due to its protection from the elements.

It also offers a public space different to that of the nearby park and enhances the connections among residents due to frequent use. The core of the complex is described as an "outdoor living room" where the design achieves its quality by using street furniture, decorative walls, ceiling and floor finishes, large pots with plants, and even chandeliers.

The spatial organization of the towers is planned so that views from the site to the nearby park are not obstructed. This, along with a permeable roof plate, permits vibrant sun exposure to all units.

The permeable pattern on the roof is also used on the ground floor, creating two courtyards with vivid light that are also the access point to the first-level units.

The re-design of Buurt 9 not only creates a compact complex but also a community among the residents due to its intriguing play with volume, spatial organization, and the surrounding context, all the while, maintaining the site's historical integrity.

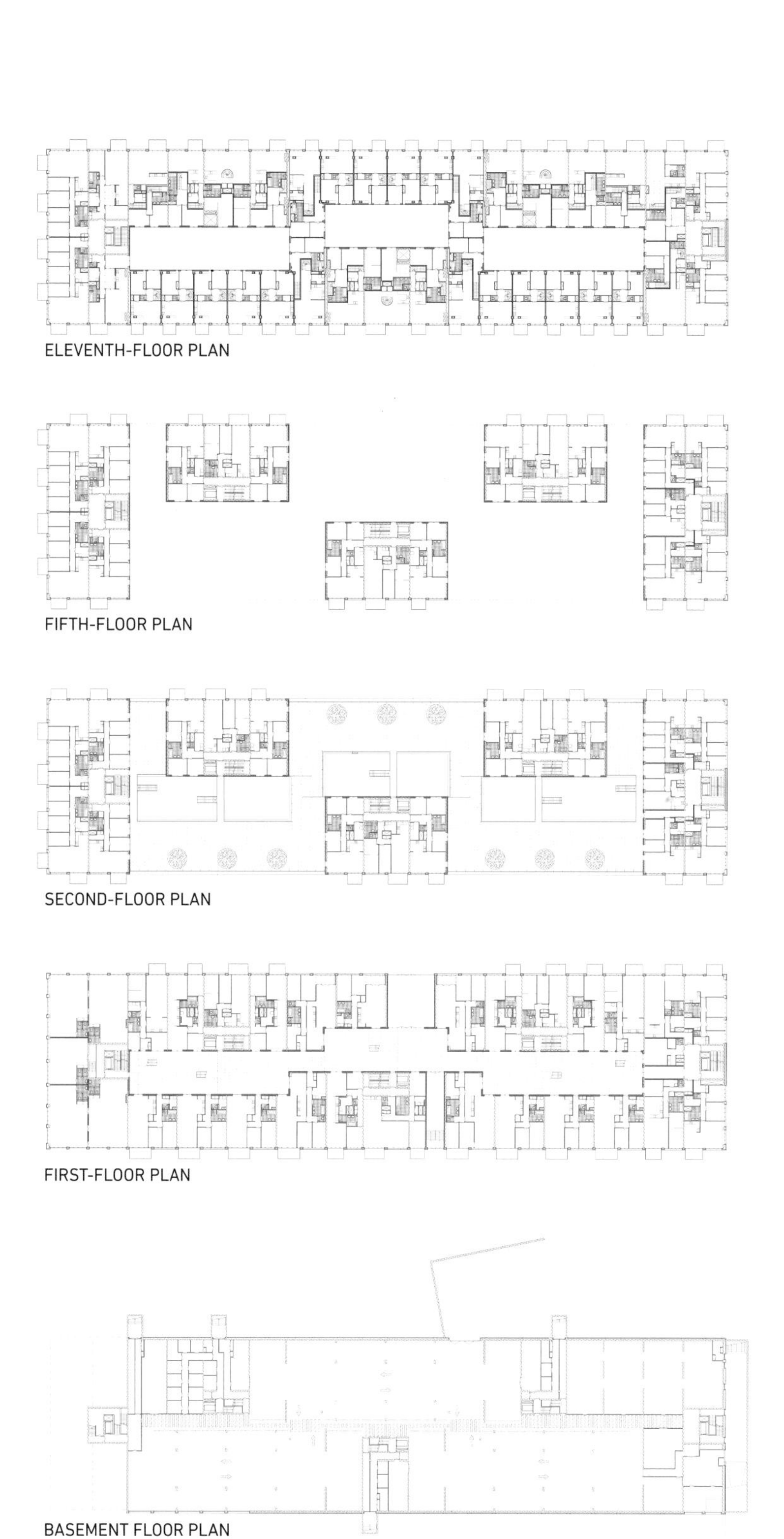
ELEVENTH-FLOOR PLAN
FIFTH-FLOOR PLAN
SECOND-FLOOR PLAN
FIRST-FLOOR PLAN
BASEMENT FLOOR PLAN

欧塞奇庭院社区

Osage Courts

Denver, Colorado, United States

Van Meter Williams Pollack LLP

Osage Courts is a multi-family apartment community near the La Alma/Lincoln Park neighborhood in Denver, Colorado. Based on public transit convenience, this development includes 185 rental units walking distance from the Osage rail station. This compact development was built on top of a parking structure, which allows three landscaped spaces to act as podiums and create semi-private courtyards for the residents, augmenting the sense of place.

The designers of the project included housing units for families and a surrounding environment for pedestrians to enhance social interaction. The architects' challenge was to orient the units to benefit from an easily accessible public transit system in a neighborhood with historic single-family homes. To overcome this challenge, the design introduced a very compact layout while still maintaining the feeling of a neighborhood due to the imaginative use of landscaping. With a constrained site such as a parking zone with a total area of 2.3 acres (0.93 hectares), an efficient density of 82 units per 1 acre (0.4 hectares) was implemented in the planning process.

Furthermore, to increase the sense of a community dwelling, the interior units of each of the three apartment groups face the pedestrian courtyards instead of having a view of the nearby streets. The units have also reached LEED-H Silver certification in sustainable and economically efficient design.

To provide a feeling of detachment from urban discomforts, an important aspect of the development was that all parking was hidden from the view of residents either by garages or carports. To further enhance this feeling, planters, porches, decks, a surrounding strip of mature trees, and a detached sidewalk for pedestrians were included.

Although area and site limitations presented challenges in the initial planning process, it did not hinder the designers' scheme to create a compact community that offers a large number of apartment units yet still embraces the feeling of walkability, easy circulation, and a detached green urban settlement.

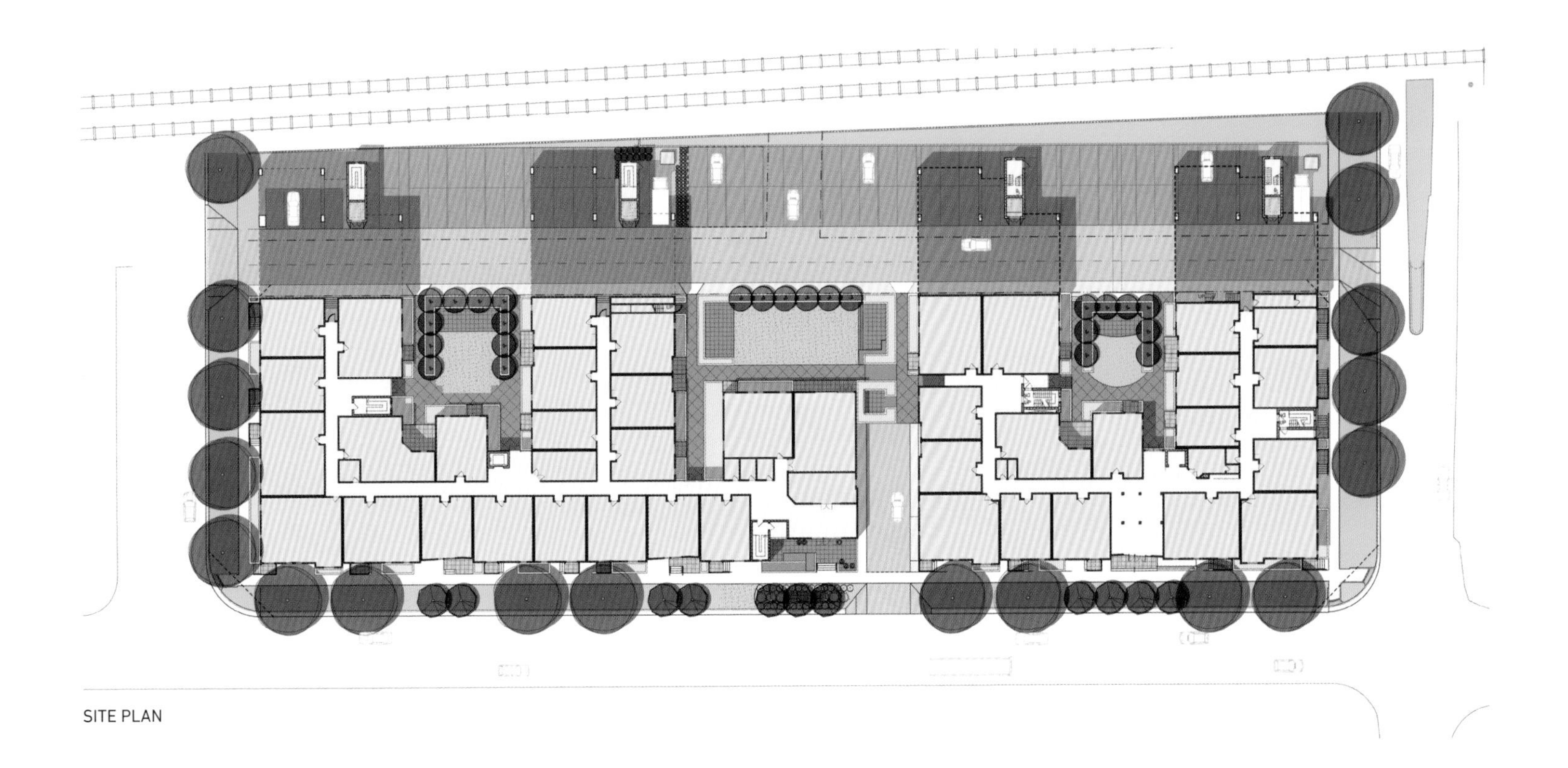

SITE PLAN

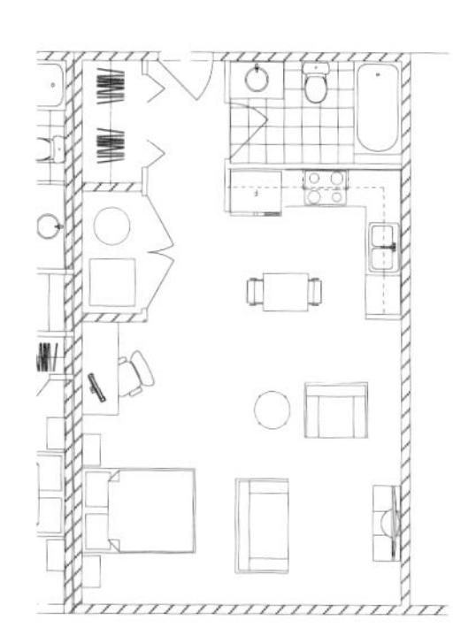

Studio – 517 square feet
(48 square meters)
9 total units

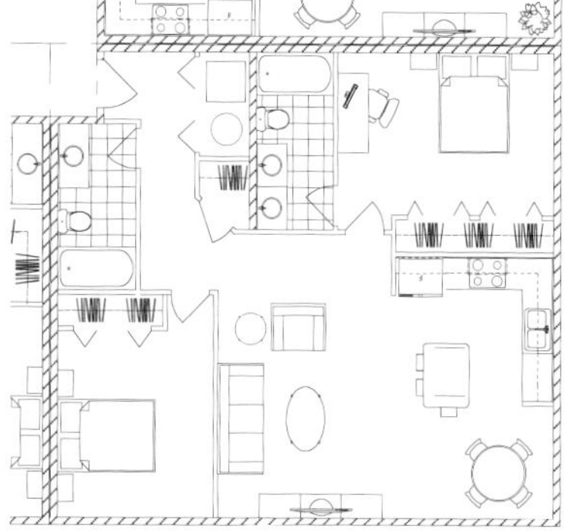

Corner unit – 917 square feet
(85 square meters)
34 total units

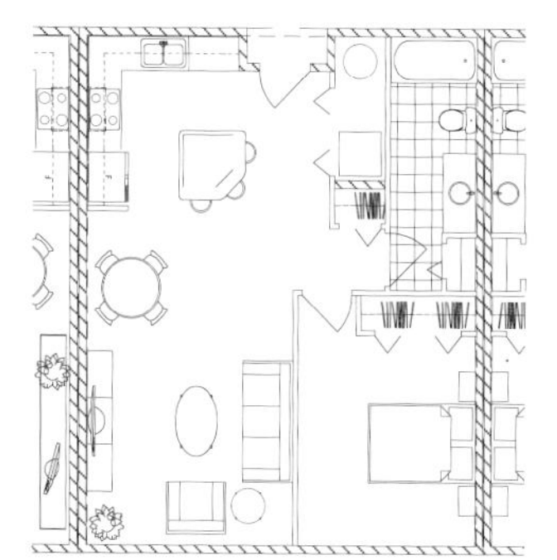

Typical one bedroom – 659 square feet
(61 square meters)
86 total units

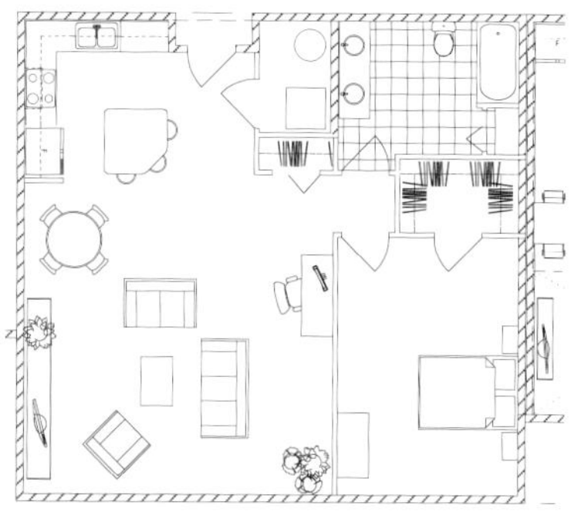

Large one bedroom – 892 square feet
(83 square meters)
7 total units

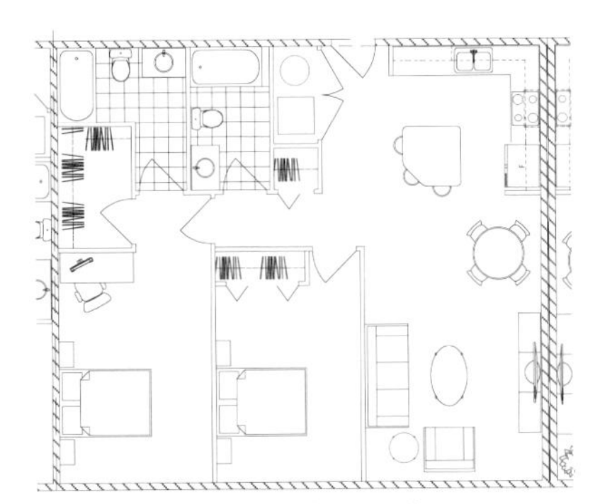

Small two bedroom – 962 square feet
(89 square meters)
9 total units

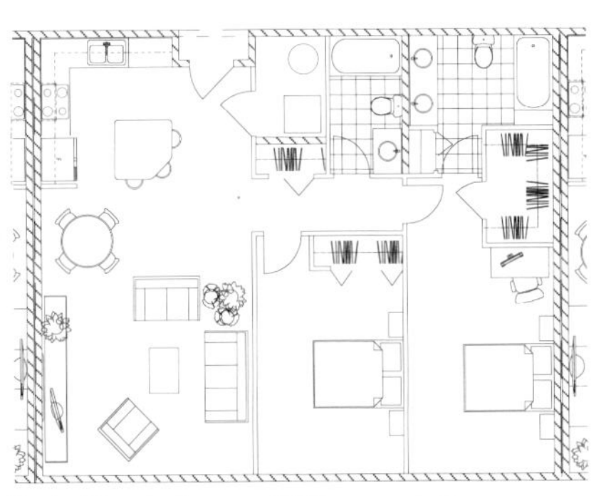

Typical two bedroom – 1007 square feet
(94 square meters)
16 total units

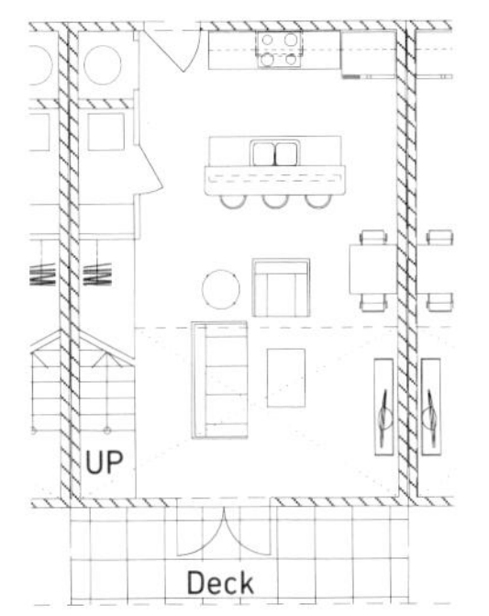

Loft main level – 409 square feet
(38 square meters)

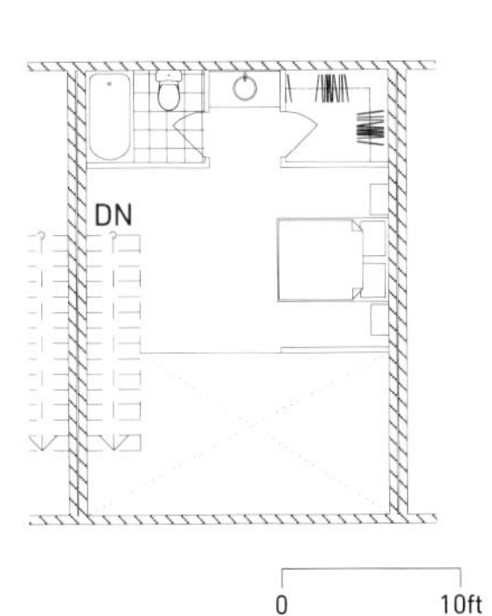

Loft upper level – 260 square feet
(24 square meters)
16 total units

第六章：新旧混合

WEAVING OLD AND NEW

Architect Carl Elefante suggests that the "greenest" building is the one that is already built (British Columbia 2012). This statement summarizes the notion that heritage conservation and sustainability are intricately linked. Indeed, preserving and reusing old buildings and revitalizing communities benefits the environment more than building new neighborhoods. Rehabilitating a building consumes less energy than constructing a new one, with the end result of reducing greenhouse gas emissions. Building in an existing area also prevents urban sprawl. This chapter explores the evolution of heritage conservation, examines the advantages of infill housing, and outlines their design principles.

Heritage conservation evolved from preserving archaeological and architectural monuments to include "intangible heritage" such as language, music, costumes, and rituals. Preservation is a dynamic process in which communities actively participate in deciding which elements of their cultural past should be kept for future generations (GHF 2012). In the decades that followed, the practice began to pay attention to cities and cultural landscapes (Turner 2013).

It is important for communities to conserve old buildings because they define the character of places, and may in turn, satisfy the psychological needs of citizens who search for meaning and identity in their surroundings. Familiar and old landscapes, therefore, have a value; they can be a source of comfort and help cope with change (Snyder 2008). People will feel a greater sense of connection to the place where they live through a shared history. It is also argued that cities that belong to their citizens become an economic, social, and cultural asset that is passed on to future generations (Council of Europe 2008).

Contemporary communities are involved in heritage conservation through their urban planning policies. Planning models such as the "Communicative Model" allowed planners to explore the sense-of-place component of conservation planning and have public input included in the decision-making process. As a result, the goals of preservationists, conservationists, and urban planners or developers eventually will merge (Snyder 2008).

The reciprocity between these disciplines also complements contemporary sustainable development philosophy. A commitment to sustainability within a community requires the use of leftover space in neighborhoods to prevent urban sprawl. Indeed, a controlled densification must be introduced by infill projects in communities, while ensuring that these projects do not disturb the quality of life in existing neighborhoods.

Infill housing design consists of finding a new purpose for vacant tracts of land and for properties that must be converted because their current designated uses are no longer viable (City of Pointe-Claire 2010). Schools or churches that have been closed down are examples of buildings that must be given a new function. There are several reasons why infill housing projects are desirable in a dense neighborhood context.

First, it is inadvisable to delay the attribution of a new function to unused buildings, since the community is still incurring operations and maintenance costs for infrastructure servicing for these buildings, even when buildings are not in use. Examples of public infrastructures that must be maintained regularly are waterworks, sewers, and streets. Next, land that is left vacant with no justification in a highly urbanized area contributes to urban sprawl. Farming or natural areas will become urbanized, increasing economic and environmental costs due to road building and traffic. The expansion of suburbia and environmental degradation can therefore, in part, be avoided by utilizing vacant land in a neighborhood, and thus contribute to sustainability by densification.

Aside from the environmental benefits, infill housing projects may also contribute to revitalizing run-down neighborhoods. In addition, because of the aging of society some argue that an increase in the number of single-person households will lead to a change in housing and neighborhood types. Infill housing is said to meet the needs of an aging population. Neighborhoods that are close to transit systems, stores, and restaurants will most likely be favored.

Infill housing may also present economic benefits for a community. They are often smaller in size and affordable, which appeals to singles, young couples, and empty nesters. Infill houses are also appropriate for people who do not own cars and want to reduce transit time by living near amenities. Furthermore, filling up leftover lots or replacing old buildings improves the aesthetic quality of a community and increases the property value of adjacent homes. There are, therefore, many economic benefits to infill housing for neighborhoods.

Townhouses are particularly appropriate for infill housing projects because they can accommodate a large number of people in a small area. As a first step, an appropriate site must be selected. There is a need to determine whether a community can accommodate infill housing as well. The site of interest must be closely examined, since infill projects must be woven into a social, cultural, and geographical context. Before planning can start, an infill housing project must first be approved by a local government.

In general, infill housing projects are easily accepted in neighborhoods that are already densely populated. For example, a high-density townhouse project will be out of place in a low-density neighborhood. Information about the site must include history, topography, soil type, drainage, and any existing infrastructure. The size of the site is also important since smaller sites are less suited for accommodating long rows of houses than their larger counterparts. Once the site has been selected, the surrounding buildings, green areas, and available parking must be considered in the design process. Existing structures will necessarily influence the design of infill housing because such structures may create unwanted wind tunnels or cast shadows. A new house should not decrease the amount of sunlight that an adjacent house already receives. The location of windows in existing houses must therefore be considered; light must still be able to reach these windows.

Close attention must also be paid to parking since infill houses are a high-density dwelling type. There is a risk of overwhelming the streetscape and the neighborhood if too many new parking spaces need to be created. Cars belonging to the incoming residents that are not kept off the streets may disrupt the traffic on an already established road, for example. One solution would be to provide private garages in front or behind the house, or in common garages above or below ground level. Other options are shared driveways or parking under rear balconies. There is also the cash-in-lieu method, which exempts homeowners from having on-lot parking spaces in exchange for paying the approximate cost of creating parking spaces in the city. This option should only be considered if parking spaces are available nearby. It is necessary even then to realize that such an option may create spill-over parking in other neighborhoods.

In general, an infill housing project must fit into the urban fabric and the architecture of neighboring buildings (Friedman 2011). The project must be woven well into its context to be accepted by the residents. To achieve that, the overall scale and mass of a neighborhood must be respected. Setbacks should also echo those of existing houses, since differences in setbacks are easily noticeable. The new house should furthermore incorporate similar stylistic elements to those found on other homes in the neighborhood. These elements should not be replicated, but reinterpreted instead, to create a relationship between the new and the old. Ensuring the continuation of streets and sidewalks will also contribute to the new dwelling fitting in, since it will reduce the impact of additional traffic caused by the new infill homes. Lastly, the arrangement of a house on its site needs to be determined based on a lot's size, shape, location with respect to existing roads, lanes, parks, and type of parking provided.

Infill housing design is greatly concerned with the integration of a project into a particular context. When infill homes are combined, they will create dense communities where old and new are mixed. Such neighborhoods help protect the environment reducing the need for transportation and by postponing the urbanization of green areas.

There is commonly an opportunity for an inversion of the relationship between private and public space by reorienting infill houses. These homes can help redefine the street and open space network in addition to providing dwellings. One row of homes may face an existing street, while the other would face a private court with a designated use for shared parking. Two rows of houses may also be placed perpendicular to the existing street and front onto a central green courtyard (Friedman 2011).

Sustainability has become a principal concern in today's society. Heritage conservation and sustainable development are linked because they deal with similar concerns. Infill housing projects are an example of conservation and sustainability working hand-in-hand, where old and new mix. While planning policies aim to address current issues, it is important to note that the built environment is never static. The challenge faced by future generations is to adapt to new realities and environmental conditions as they occur.

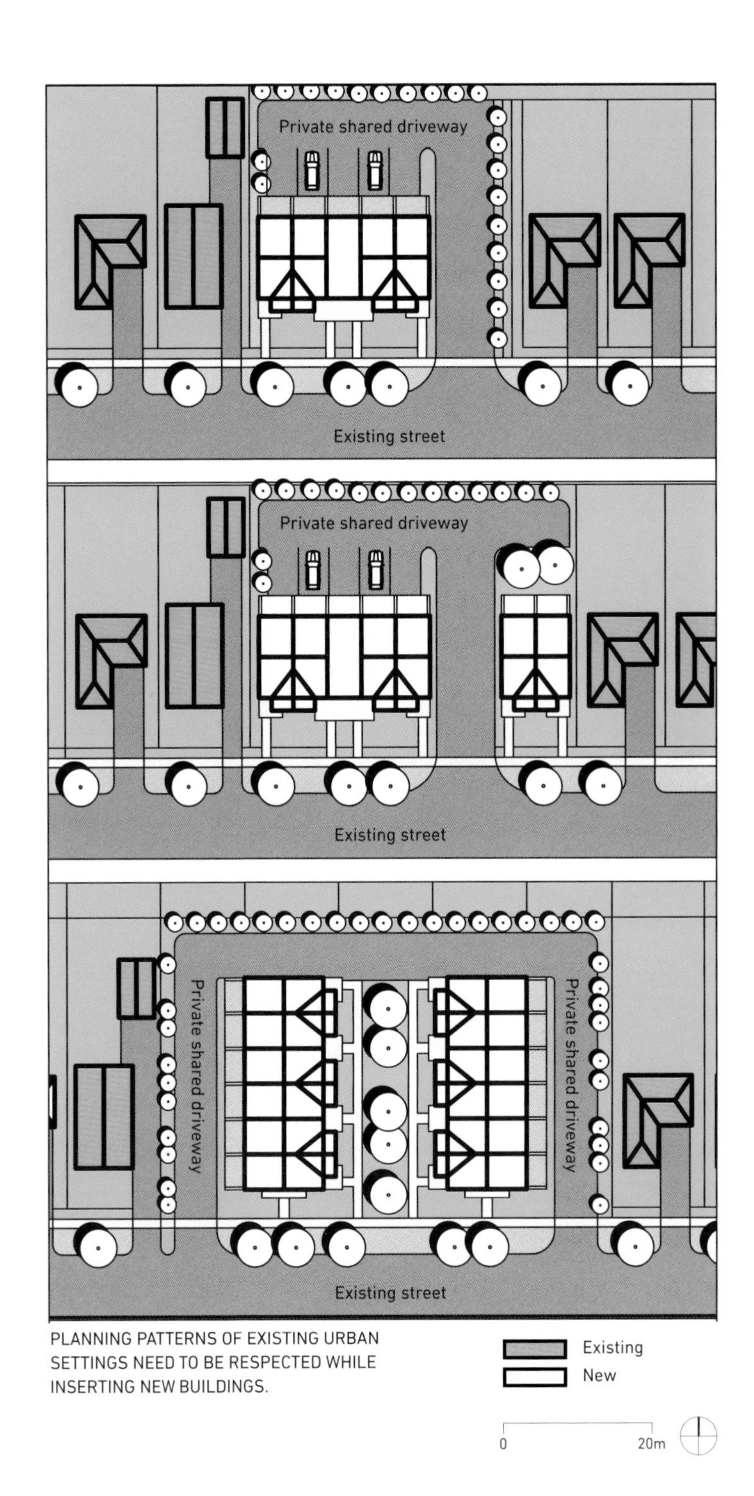

PLANNING PATTERNS OF EXISTING URBAN SETTINGS NEED TO BE RESPECTED WHILE INSERTING NEW BUILDINGS.

艺术风景韦奇伍德谷仓社区

Artscape Wychwood Barns

Toronto, Ontario, Canada

DTAH

Designed by DTAH, the Artscape Wychwood Barns are located near St. Clair and Christie Streets in mid-town Toronto, on top of a hill that marks the ancient shoreline of Lake Iroquois and immediately adjacent to Wychwood Park, one of Toronto's earliest planned residential communities. The project transformed the early 20th-century streetcar maintenance facility into a vibrant community hub focused on the arts and environment, while preserving the historic authenticity of the place.

The 1.23-acre (0.5-hectare) site is owned by the City of Toronto, with Artscape signing a 50-year lease in exchange for funding and operating the facility. Through an innovative mix of programming and the establishment of strategic partnerships with organizations such as The Stop Community Food Centre, Artscape has transformed the site into a vibrant center for artists, environmentalists and the general public to live, work, gather, and learn.

The complex comrpises the five existing linear maintenance buildings, or "barns", which were adaptively reused to accommodate 26 artists' live-work studios, 15 artists' work studios, a community gallery, and 15 office suites for non-profit organizations. Other programs include a 5,000-square-foot (465-square-meter) commercial greenhouse, community bake oven, outdoor growing areas, and a covered street, which accommodates farmers' and artisans' markets, conferences, and events. The vision for the complex was developed through a feasibility study that included a substantial public consultation program over a number of years. This led to consensus building among the community in terms of the use of programs on the site.

4
3

Rainwater is harvested from the roof and connected to all toilets within the building as well as irrigation systems that serve the surrounding City of Toronto Park (designed by the Planning Partnership). Daylighting is available in all interior spaces supplemented with energy-efficient lighting throughout the facility. In addition, the heating system is geothermal, reducing natural gas consumption significantly.

The project has garnered distinction because of its cultural significance and has also received a LEED Gold certification. Additional recognition that Wychwood Barns has received include the Charter Award from the Congress for New Urbanism, the Design Excellence Award from the Ontario Association of Architects, the Canadian Green Building Award from Sustainable Architecture and Building Magazine, and the Peter Stokes Restoration Award.

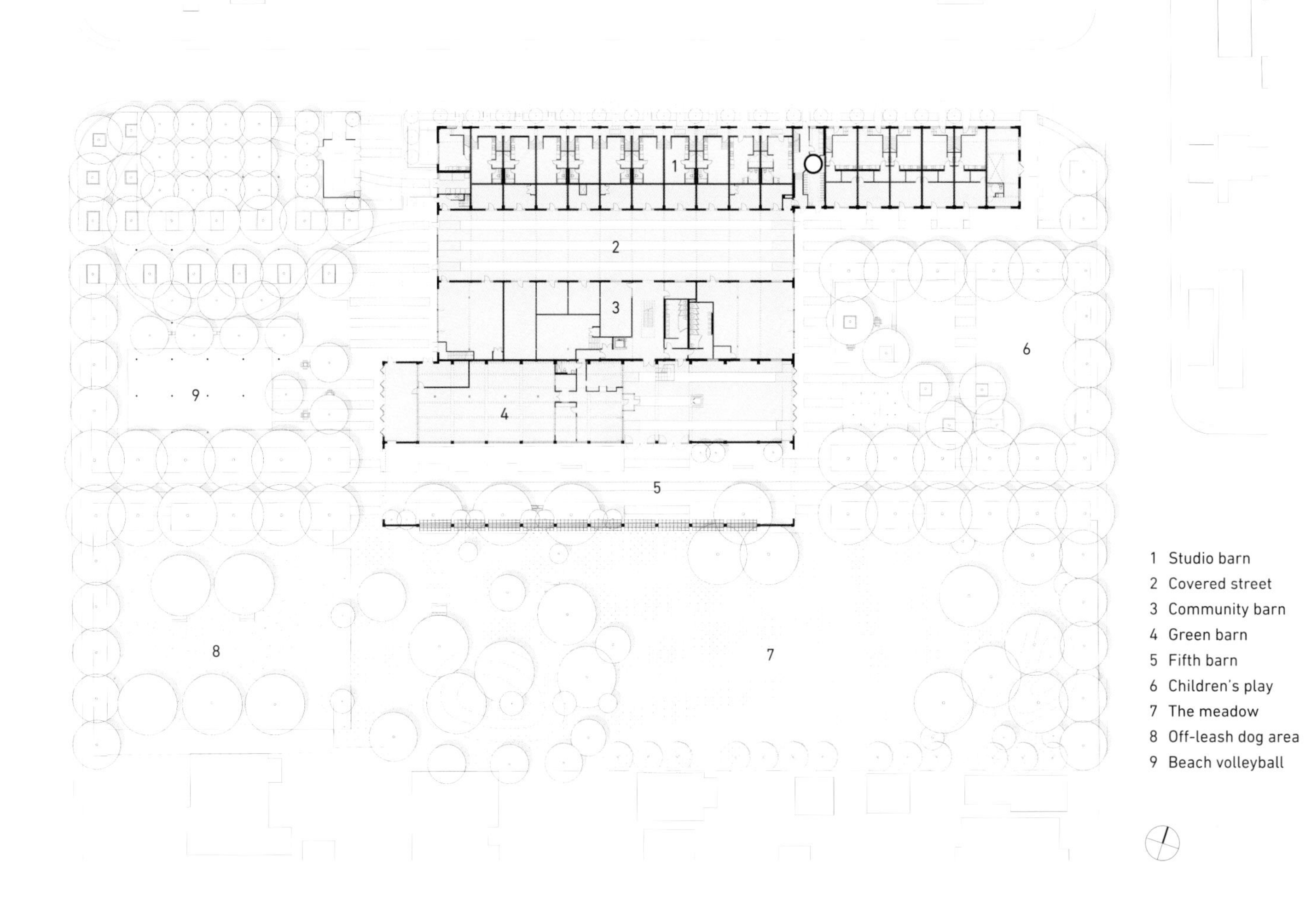
1
2
3
4
5
6
7
8
9
1 Studio barn
2 Covered street
3 Community barn
4 Green barn
5 Fifth barn
6 Children's play
7 The meadow
8 Off-leash dog area
9 Beach volleyball

SITE PLAN

WYCHWOOD BARNS
3
4

24号巷

Alley 24

Seattle, Washington, United States
NBBJ

Alley 24 was built in an old industrial neighborhood, on the same block as three early 20th-century historic brick buildings, home to Vulcan Inc. and PEMCO Insurance. For a fully integrated project, the brick façades at the north side had been incorporated into those of the residential sector of Alley 24. This created a visual exchange between the old and the new, but one that does not inhibit the clean industrial aesthetic of the neighborhood. The aesthetics of the project itself are not solely dependent upon context, but are largely influenced by the sustainable features and functions that help its users to be environmentally conscious without sacrificing comfort.

With a site area of 8.12 acres (3.28 hectares), the project takes into consideration the variety of residents present in the neighborhood and does its best to create a harmonious community. Alley 24 is composed of sections of housing, office space, and ground-level retail with a range of one to six stories. The 172 residential units vary from studios to two-room apartment units for a mixed income. The hope is to attract designers and other professionals to this dynamic community.

In addition to integration of the massing into existing structures, Alley 24, like its name, incorporates an existing alleyway that divides the lot into two, the residential and the office spaces. A second alley was added to the site that is parallel to the original and breaks up the massing further. The designers added, with intricate planning, spaces for foot traffic, gardens, and multiple entrances and retail spaces along these paths.

SITE PLAN

P ←
APARTMENT
LEASING
OFFICE

Not only was Alley 24 meant to be woven into its immediate urban, communal, and historic fabric, it was also designed to be fully integrated with sustainable technology. It includes hybrid ventilation systems, optimization of day lighting, and under-floor air distribution systems. There is even a built-in automated system that notifies tenants when the external temperature range is between 62°F to 72°F (16.7°C to 22.2°C) to turn off the mechanical air conditioning and open the windows. Other sustainable strategies include façade optimization to admit daylight and moderated solar heat gain. External shading systems, such as aluminum sunscreens and automated blinds, respond to changing environmental conditions and are used in combination with electric lights that turn off automatically when natural lighting reaches a certain level.

Alley 24 was one of the first American projects to be awarded the LEED certification for mixed-used projects. The project ventures beyond energy saving techniques to integrated green design.

SECTION

拉格公寓
Rag Flats

Philadelphia, Pennsylvania, United States
Onion Flats

Located on a former industrial site, Rag Flats is a collective design project and exploration into the re-contextualization of the site from its factory origins into a residential garden community. In addition to responding to a set of problems, the project is more of a method of design that is collective, creative, and cohesive, and explores relationships between density, intimacy, and privacy in a community. It also provoked a dialogue around how adaptive reuse projects can build upon the characteristics that are vital to a neighborhood, thereby enriching an existing community.

The project was designed to be incorporated into the urban fabric in a way that helped it blend in, but at the same time allowed it to have its own original identity. In Philadelphia's Fishtown neighborhood, there is a myriad of different building types and styles, dominated mostly by brick and stone row houses. Rag Flats was designed to be of the same scale and feeling as the rest of the buildings in the neighborhood, yet the use of its unconventional materials made it stand out. Steel, aluminum, wooden slats, and painted stucco were employed to help the project attain a character of its own.

Four distinct but familiar Philadelphia typologies in a total of 11 residential units can be found within the complex; the row house, the trinity, the loft, and the pavilion. They were transformed from the former two-story rag factory structures, which has been used as dump sites and were on the verge of collapse. The factory had previously been a central point of the community and many nearby residents had once been employed there. That space had been used to explore ideas about what makes a city, but most of all how to construct a sustainable community.

Sustainable features have been integrated into the design. The courtyard utilizes permeable pavement to filter stormwater and reduce runoff. A 6000-gallon (22.7-cubic-meter) rainwater cistern is used for the water collected on the rooftop, which also contains an 8-inch (20.8-centimeter) deep green roof to reduce solar gain and a 32-kilowatt photovoltaic system to help reduce the dependence of residents on electricity from the city grid. These green-conscious design methods not only help the environment but also help to construct a place on the shoulders of its industrial past.

伊顿天然社区
Eden Bio

Paris, France
Maison Édouard François

The Eden Bio project, which is also known as the Villas des Vignoles, simultaneously addresses issues of community, urbanism, and architecture due to the large site area it occupies. It explores the densification of a typical suburban block in Paris and turns it into a tight-knit community. The project has been nominated for the Silver T-square and the Mies van der Rohe awards.

With a total site area of 1.90 acres (0.77 hectares), three ideas were the guiding force behind the design by Maison Édouard François. The first and foremost is that it was crucial to respect the surrounding area with its wealth of visual history and culture. The existing buildings ranged in height from low to tall, and the alleyways were long and narrow—a clue to the site's agricultural history. They give the plots a complex spatial definition and variety. Once this idea was established, it was evident that the building program needed to limit building on street alignments, to preserve the existing suburban alignments, and to respect the alleyways so they could function as connecting paths among the residents. A long low-rise building was placed at the very center of the complex, surrounded by extensive vegetation and small authentic townhouses with wood, cinder blocks, mechanical tiles, zinc, and exposed-concrete façades.

The second idea revolved around access the project. The block is a labyrinth of stairs and lattices. There are no connecting corridors in the interior, only individual entrances. These entrances express the individual identities of the residents and guide circulation throughout the complex. The external stairs rise above and beyond the planted façades, and lead to two units on each level. To allow the natural surroundings to seep through into the living spaces, each apartment unit contains windows on either side of the building.

The third idea is about allowing nature to inhabit the nooks and crannies of the development, forming a "village-like" composition that is varied and far from monotonous. In other words, the open spaces within Eden Bio are not simply planted gardens, but abandoned spaces that experience plant growth over time. To achieve this, the original soil of the site was substituted by a deep organic soil.

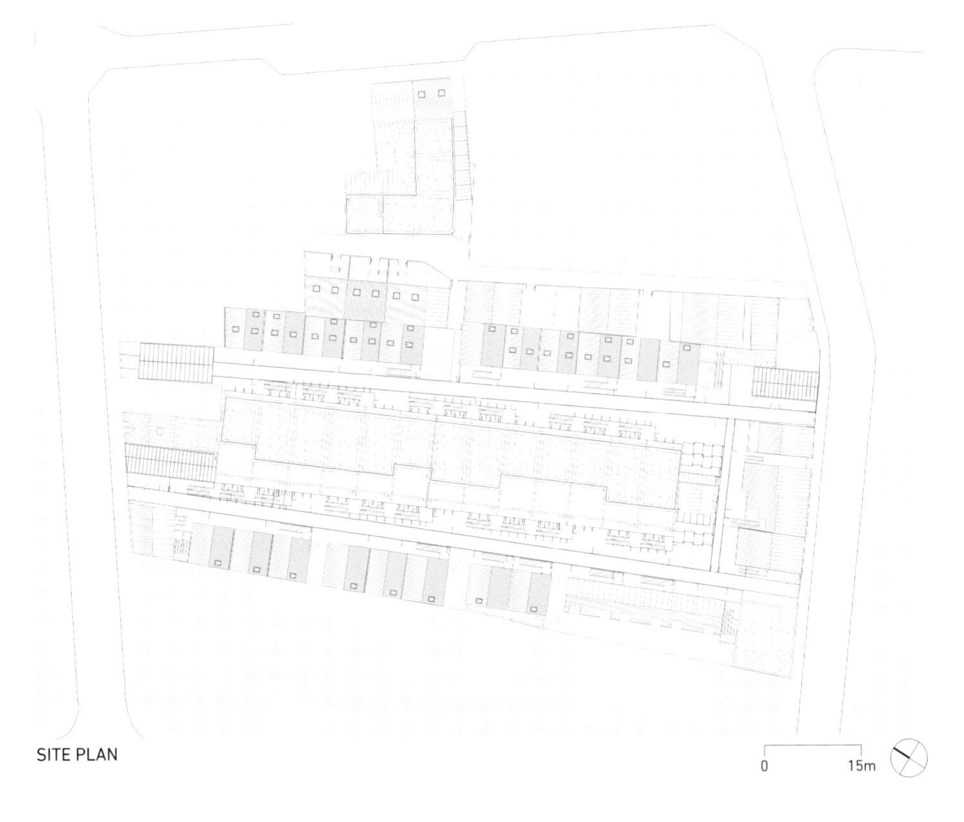

SITE PLAN

卡恩·里巴斯工厂改造项目

Can Ribas

La Soledat, Palma de Mallorca, Spain

Jaime J. Ferrer Forés

In 1851 the Can Ribas factory in La Soledat, Palma de Mallorca, Spain was a place where woolen blankets were produced. As the factory and population of workers grew, it became a complex consisting of several buildings covering an area of 1.30 acres (0.52 hectares). The existing housing was built for the employees around the factory, marking the beginning of an urban neighborhood. Later on it became one of Palma de Mallorca's main industrial working-class districts.

The closing of the factory in the 1960s led to the economic descent and social marginalization of a neighborhood that had lasted for over 50 years. It also led to the deterioration of the dwellings, streets, public spaces, greenery, and commercial properties. The Can Ribas factory had become a sort of wall, dividing the community into two. In the early 2000s, the Department of Housing in the Palma de Mallorca City Council offered to re-design the core of the community. One of the urban planning strategies was to elongate the eastern street Carrer de Brotad to the northern street of Carrer de Manacor in order to connect La Soledat to the more modern northern urban communities. Furthermore, Carrer de Brotad also needed to be connected to the eastern district of Poligon de Llevant. Another strategy was to construct 160 social-housing rental units. To accomplish this, tearing down the walls around the perimeter of the factory complex was necessary. The reconstruction of the existing buildings and the introduction of green areas was also needed so that the project could be open to the public. The goal was to enhance the historical aspects while repairing the factory to create a well-organized neighborhood.

The residential areas have been constructed on the southern side of a newly developed public square, with the old factory chimney kept as a symbol of heritage preservation. The northern part of the square is lined with façades of existing buildings, which have undergone renovation. These buildings were planned to contain two public facilities, an Interpretation Center and a Cultural Center. One of the other two newly renovated buildings that face Carrer de Brotad on the southern part of the public square, contains a large porch for public entertainment and events.

With the development of 160 units, this urban community uniquely blends historical qualities with its new functional goals. Using circulation to connect with rest of the city stimulates urban revival. Providing green areas and re-structuring existing public spaces to give it a more dynamic and contemporary feeling further augments the status of a community that is set to leave its mark for generations to come.

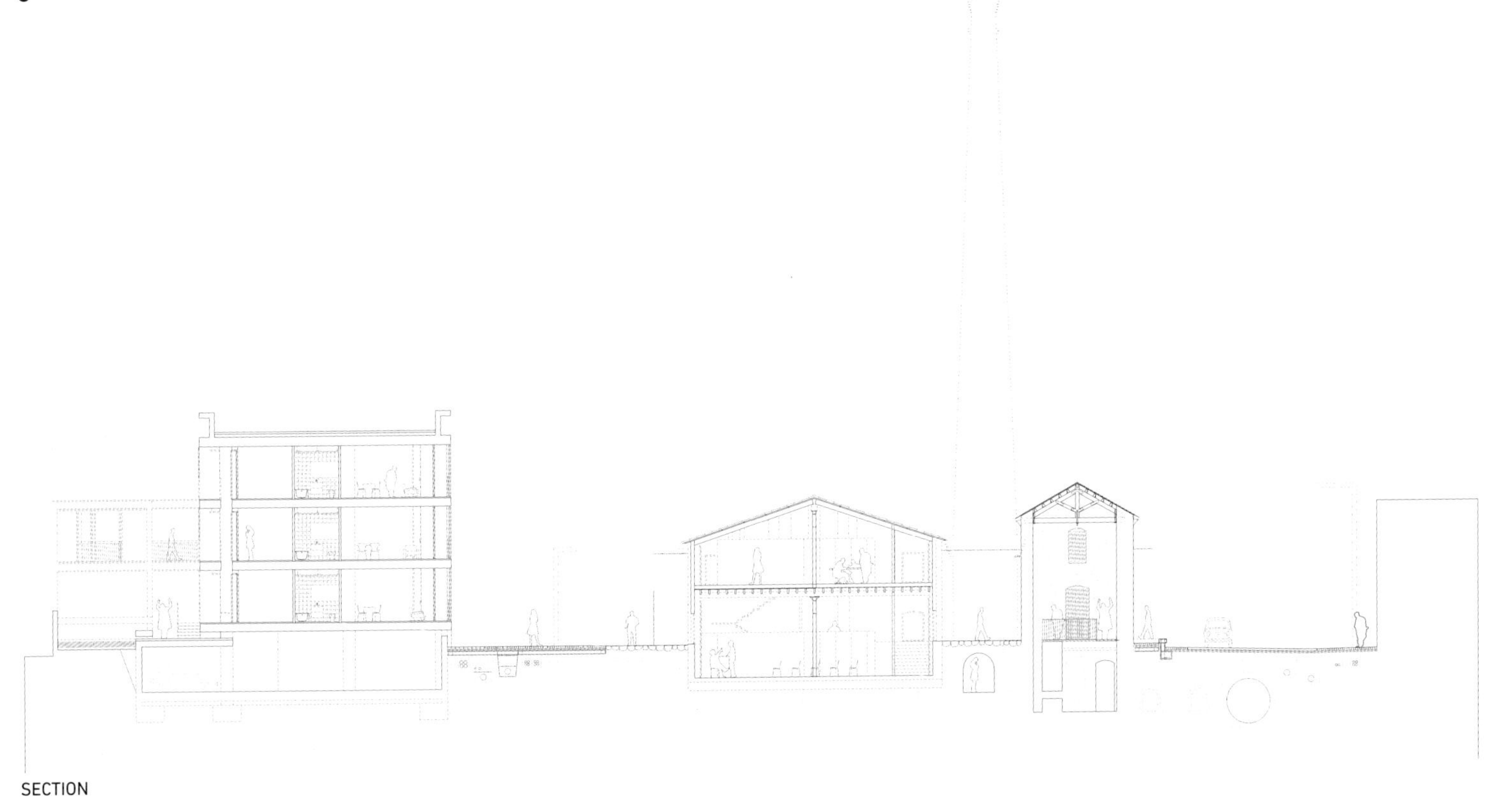

SECTION

第七章：可食用景观

EDIBLE LANDSCAPES

7

The international food trade has increased by 184 percent between 1968 and 1998 (NRDC 2007). This is not without consequence. Large amounts of CO_2 were emitted in the process of transporting this food. For instance, a refrigerated container ship from Costa Rica travels over 1802 miles (2900 kilometers) to deliver fruit to US ports, resulting in 96.82 pounds (43.92 kilograms) of CO_2 emitted per ton of fruit (Paster 2011). As a result, the need to establish more sustainable methods of food production is evident. Recent initiatives include community gardens where people can grow food locally. This chapter explores the practice of urban agriculture within the neighborhood setting.

Urban agriculture is defined as the "cultivation of plants, medicinal and aromatic herbs, fruit trees, and the raising of animals in cities" (Bhatt 2005). Locations in neighborhoods may be "the homestead (on-plot), or on land away from the residence (off-plot), on private land (owned, leased), or on public land (parks, conservation areas, along roads, streams and railways), or semi-public land such as grounds of schools or hospitals" (RUAF 2013).

Urban agriculture may also have a favorable social effect on a neighborhood. Engaging a community in local food production improves its health and provides an opportunity to earn additional income. People with fewer employment opportunities may also be able to become entrepreneurs. Moreover, urban farming is often undertaken through a community organization. As a result, a greater sense of solidarity will be developed; the community will have a shared concern for the success of the enterprise and the fruits of its labor (Smit 2001). Communities that participate in urban farming also have higher levels of social interaction and better security in general, since activities take place outdoors.

Urban agriculture also helps to landscape neighborhoods by turning unsightly lots into neatly cultivated areas. Green spaces help improve mental health and have therapeutic benefits derived from participating in planting and agricultural activities. Another health benefit of edible landscapes is the improvement in air quality. Packed soil is converted to loose, regenerated soil by returning organic material and microbes through the reuse of solid waste and wastewater. Loose soil is desirable because it cleanses water and promotes the plant growth that cleans the air (Smit 2001). Plant and related insect and animal life are consequently promoted, and the increase in vegetation helps improve air quality. The amount of dust in the air is reduced and pollutants are absorbed through the foliage.

Other advantages of urban farming are its food safety and nutritional benefits. The quantity, quality, regularity, and nutritional balance of people's diet have a great influence on their health and wellbeing. Furthermore, edible landscaping helps reduce hunger by providing easier access to food. Fruits and vegetables, for example, can be procured through home production. This is especially beneficial for families on a lower income. Nutrition will also be improved since produce farmed locally is fresher than imported food (Smit 2001).

Local production brings food closer to consumers and is also site-specific, since certain types of plants are more suited to the climate and the environment of certain regions. Often times imported fruits, vegetables, and meat will spoil during transportation or from improper storage. In addition to being fresher, local food farming also reduces the amount of food waste, contributing to the protection of the environment. In brief, the benefits offered by urban agriculture are both quantitative and qualitative—increasing food quantities reduces hunger, while improving food quality fosters better health (Smit 2001).

Urban agriculture has considerable economic benefits for a community as well. It can be started on a small scale with little monetary investment, limited technical knowledge, or skill. Urban farming presents the opportunity for a secondary income for private and public organizations. Economic activity in related industries will increase as well, including those that supply agricultural accessories, storage, transportation, canning, marketing and food processing (Smit 2001). Moreover, prices might be reduced by local production since the food will pass through fewer channels. By producing food locally, families may reduce the amount of money spent on food, investing the savings in health care or education.

Local food consumption reduces "food miles" defined as the distance food travels from where it is grown to where it is ultimately purchased or consumed (NRDC 2007). Imports by plane generate more greenhouse gas emissions than imports by ship. In 2005, the import of fruits, nuts, and vegetables from California by airplane released more than 70,000 tons of CO_2, which is equivalent to more than 12,000 cars on the road (NRDC 2007). Moreover, producing food in a neighborhood helps conserve the place's biodiversity. Urban farming can produce [five to ten] times as much per acre as rural farming (Smit 2001). It can be said that a community garden may conserve an area ten times "as large in a remote rain forest or mountain range" (Smit 2001). There is, therefore, a significant environmental incentive to producing food locally—it is a highly sustainable process.

Typical products raised in urban gardens include grains, roots, vegetables, mushrooms, fruits, poultry, herbs, and ornamental plants. Individual motivations to cultivate food locally vary: urban farmers may seek nourishment, extra income, or recreation. Community gardens allow people who live in apartments or multiplexes, or who do not have access to a home garden to cultivate their own land. Regardless of the reason for cultivating food, however, edible landscaping is considered a sustainable practice because it encourages local production and consumption of food. What ensues is a reduction in energy input and output caused by industrial food production, which contributes to protecting the environment.

The organization and planning of an urban garden is crucial to its success. The garden must be integrated well into its community. A low-maintenance and high-yield garden can be created from strategically making use of natural factors such as sunlight and water flow to their best advantage. Rainwater for example can be collected, distributed, and drained as needed if an effective system is designed. Rainwater from roofs can be redirected into plastic collection barrels, swales, berms, or drywells for later usage and automated garden watering. Deep watering encourages the roots of plants to grow downwards.

The sun path must also be analyzed to determine the sunny and shaded areas in a garden design, which will influence the placement of different plant species. Either a lack or an excess of sunlight or water will result in a lower return. Several things need to be considered during the planning of an urban garden, among them layout, existing structures, lawn size, orientation, topography, climate, and choice of plants. Soil is also important, as the absorption, depth, richness and water-retaining capacity of soil are highly significant for the successful growth of a plant. Different plants have different needs so each plant's needs must be considered individually. Plants with similar water and sun needs may be placed together, but a 2.5- to 2.98-foot (0.76- to 0.91-meter) wide path should be left open for easy access to vegetable crops. At times though, vegetables and herbs can be combined into the same zone to encourage pollination, but trees should remain separate because of underground root space requirements and competition for sunlight. Different plants also require varying soil acidity levels; most plants thrive in neutral soil, while peppers and potatoes require more acidic soil.

Different scales exist for the practice of urban agriculture within a neighborhood. Individuals may begin urban farming in their backyard or front lawn. Some plants such as basil can even be grown inside the house. Larger-scale urban farms include gardens shared by schools, hospitals or universities. Community gardens, where individual plots are cultivated by different people, involve a greater segment of a neighborhood. Community gardens allow people who do not have a large enough plot of land the possibility to grow their own food.

Placing an urban garden in every residential cluster creates a pleasant living environment. Houses would have a view of the garden as opposed to a view of a parking lot, for example. Vertical farming is an option for households that only have a small lot or do not have one at all. Two types of vertical framing exist: vine-like plants that are supported with strings, walls, or alternatively fences where crops are layered in two or three levels. The second type is useful for plants that require large amounts of watering, as those plants can be placed on the lowest level. Similarly, those same plants should be placed at the bottom of a plot if it is slanted.

Other types of private residential gardens include greenhouses, small yards, and rooftops. Since greenhouses absorb and trap heat, the growing season can extend beyond natural cycles. Greenhouses may be constructed as part of a dwelling or as a separate unit. Yard gardens are located in private front, back, or side yards. Tool sheds, frames, and compost bins should be readily accessible in yard gardens. On the other hand, flat rooftops are ideal garden spaces because they are leftover space that is often left unused and present a large space for an urban garden. However, because many rooftops are not structurally designed to support a garden, the weight of soil, plants, and other materials and equipment must all be accounted for in the planning of an urban garden if it is to be located on a rooftop. Regardless of the scale of the farm, urban gardens often present an opportunity for composting in addition to farming. The location of the compost is flexible, but should generally be placed where the smells would not noticeable.

Crop rotation is a further consideration in the planning of an urban garden. Rural farmers change the type of produce grown in their fields every growing season to avoid overworking the soil. A similar result can be achieved by urban farmers through mixed-planting, where a variety of plants are cultivated side-by-side. Each plant requires different nutrients and deposits different minerals, and therefore a balanced soil composition will be maintained. However, it is important to remember compatibility between plant types. Trees, vegetables and herbs can be grouped into their respective zones, for example, and it is sometimes imprudent to mix types. Urban agriculture, however, is not limited to trees and vegetables—other possible products include poultry, rabbits, goats, sheep, cattle, pigs, guinea pigs, and fish (RUAF 2013).

Aquaponics is a sustainable food production system that combines a traditional aquaculture thereby raising aquatic animals such as snails, fish, crayfish or prawns in tanks with hydroponics, cultivating plants in water (UFF 2013). Small home scale units in a neighborhood are possible, and the farm can be placed on rooftops. A multitude of types of fish can be raised in aquaponic farming, such as trout, tilapia, and salmon. In these systems, the fish's water feeds the plants, and the plants cleanse the water for the fish. Fish can therefore be produced in an urban setting and can help feed a neighborhood locally.

The scale and impact of urban farming can be increased through farmers' markets where people can sell excess produce. One of the benefits of farmers' markets over grocery stores is that people can buy fresher products. In addition, health will be promoted, and farmers will have lower transportation and storage expenses.

There are many social, environmental, and economic benefits to urban agriculture. Edible landscaping is by definition sustainable, since local food production will lower the demand for imported food and reduce the energy expended in its transportation and storage.

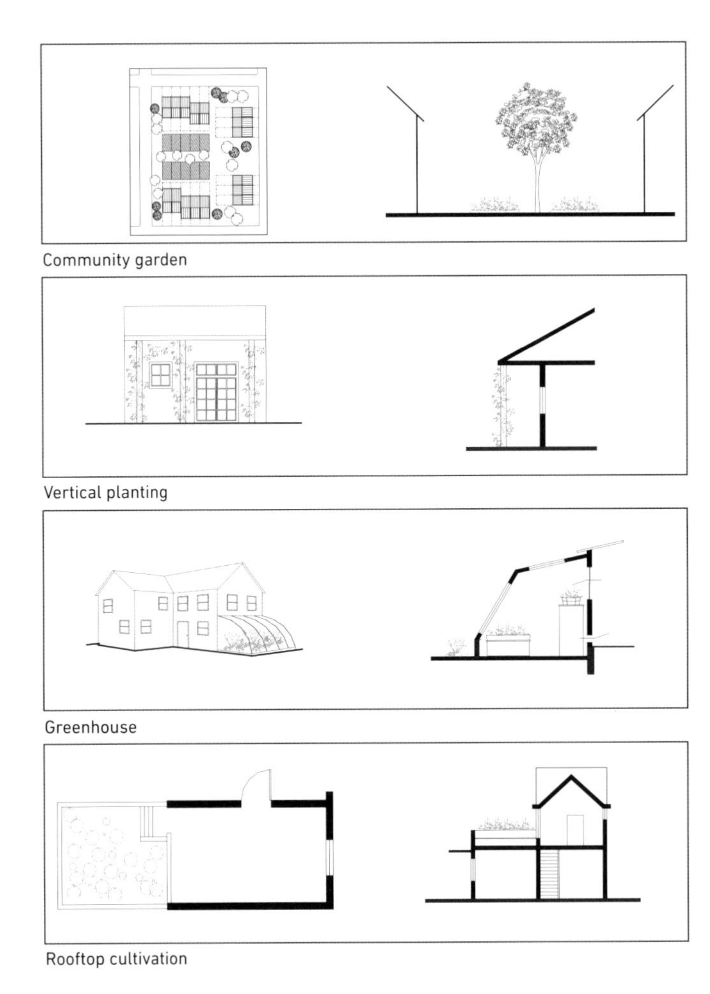

SEVERAL AVENUES OF FOOD PRODUCTION CAN BE INTEGRATED INTO COMMUNITY DESIGN OR A HOME.

青草巷公寓

Via Verde

New York City, New York, United States

Grimshaw Architects, Dattner Architects

Via Verde covers an area of 72.64 acres (29.4 hectares) and has 222 apartment units in three different building types. A 20-story tower is located in the northern part of the site, a 6 to 13-story mid-rise duplex apartment complex in the middle, and 2- to 4-story townhouses in the south. It is a public commitment by the City of New York to create the next generation of social and community housing while meeting the goals of healthy and sustainable living.

The focal point of the project is a communal garden, which guided the planning of the whole community. The garden begins as a public courtyard at ground level and then spirals up through a series of roof gardens conveniently located in the south of the site. This rising spiral of gardens creates a circulating promenade for residents. The garden makes a dramatic finish as a roof terrace where the inhabitants can enjoy a breathtaking view. These green roofs encompass the entire complex and are the very essence of Via Verde. The gardens connect the residents with the natural environment, encouraging them to cultivate their own produce, participate in recreational and social opportunities, and also provide rainwater absorption, strengthen insulation and dissipate heat. The roof gardens include evergreen trees on the third-floor roof, dwarf fruit trees on the fourth-floor roof, and vegetable gardens on the fifth. In advocating green living to all its inhabitants, each resident is given a guide with information on how to promote energy optimization and healthy living.

Via Verde was designed for mixed-use and mixed-income residents. The main access of the site leads to the residential lobbies and townhouse entrances that are located around the courtyard. Amenities such as retail, a pharmacy, a community health center, and live-work units encompass the ground floor creating a unique urban atmosphere. A fitness center is also available and located above the main entrance, which overlooks the street and courtyard. To further promote healthy lifestyles, the staircases are easily accessible and visible. The sustainable design is expressed through shared courtyards and large windows in the units that permit cross-ventilation, therefore increasing the natural ventilation of fresh air in the units and decreasing the use of an air-conditioning system. Furthermore, photovoltaic panels provide solar energy to the entire complex and are beautifully implemented to exhibit an appealing façade showcasing the exterior materials.

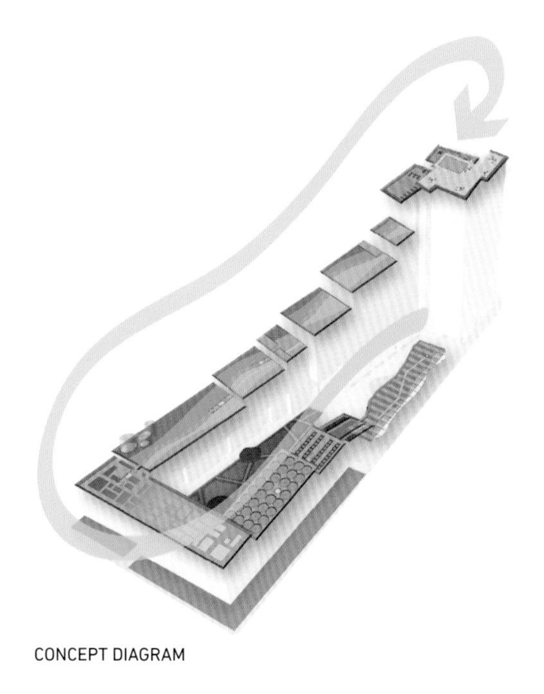

CONCEPT DIAGRAM

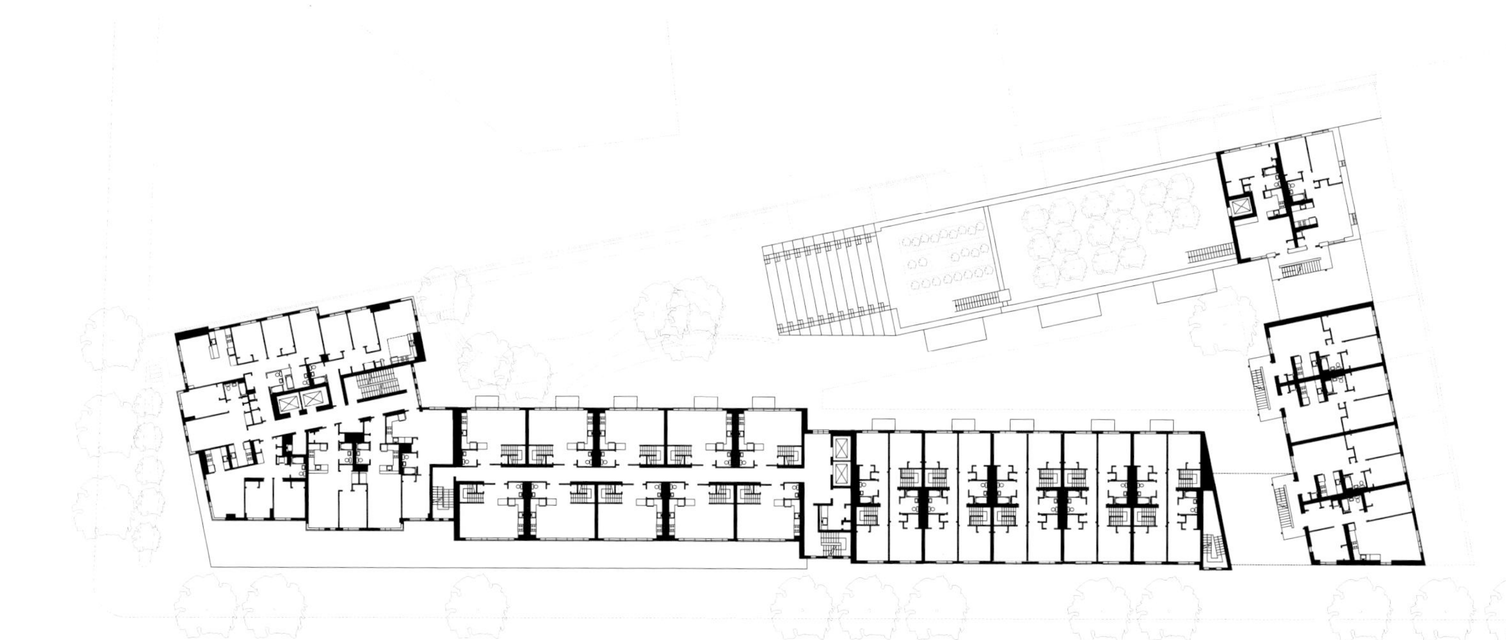

TYPICAL FLOOR PLAN

One of Via Verde's missions was to address the lack of organic food cultivation in the southern part of the Bronx in New York City. At Via Verde, residents are provided with opportunities to be fully involved in the production of growing and enjoying their own food. Furthermore, gardening also teaches children about healthy food production, safety, and alternatives—creating awareness for future generations to come.

With the development of 160 units, this urban community uniquely blends historical qualities with its new functional goals. Using circulation to connect with the rest of the city stimulates urban revival. Providing green areas and restructuring existing public spaces to give it a more dynamic and contemporary feeling further augments the status of a community that is set to leave its mark for generations to come.

沃德维尔庭院公寓

Vaudeville Court

Islington, London, United Kingdom

Levitt Bernstein

Vaudeville Court, a high-density, sustainable, and affordable housing project, is located in Islington, London. Its key concept is devoted to "productive landscapes" and it was the winning entry to an Islington Borough Council Housing open architectural competition.

The project made use of the existing residential terrace and extended the same building line and scale to a total site area of 0.32 acres (0.13 hectares). This created a community that provided residents with 13 dwellings comrpising a mixture of two-, three-, and four-bedroom family houses and apartment units. Built on an abandoned site, the Islington Borough Council chose Levitt Bernstein's scheme that exhibited the concept of "home-sown" where a space devoted to communal amenities is combined with portions of plots and linear strips of garden space for new residents and adjacent neighborhoods to grow their own fruits and vegetables. The designer's ideology of "home-sown" can be traced back to Islington's history, a prosperous site that was well known for its agriculture, dairy herds, local produce, and trade.

SITE PLAN

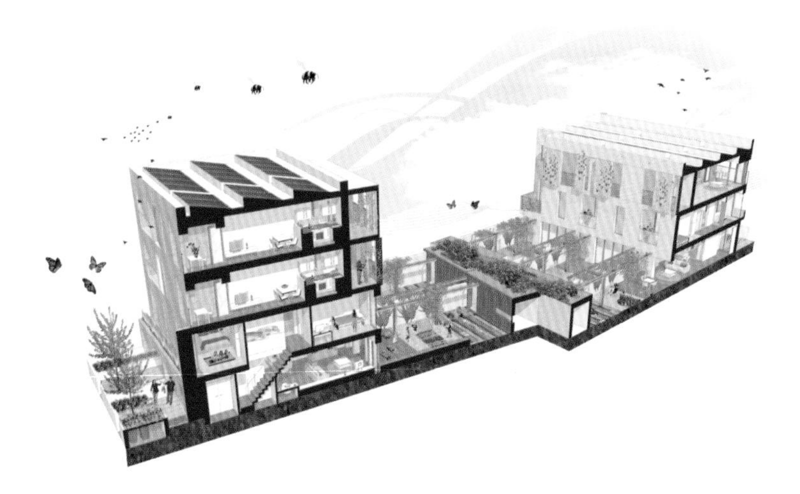

The plan for the ground-floor duplexes share the same layout; kitchens are positioned to overlook the adjacent street, the dining space is in the middle, the living space is at the back facing the courtyard, and the private bedrooms are on the floor above with their own private terrace. The living area leads to the exterior courtyard—maximizing the interior space and natural light, which combines with the exterior. The private courtyard includes external seating and a linear planter. From the planter, the space then terminates into a garden room sheltered by a trellis and sliding timber screens. This exterior layout maximizes storage space and privacy between each courtyard. To cover the private gardens on the ground floor, a network of brick panels were used, thereby encouraging more opportunities for vertical planting and the hanging of herbs and shrubs.

Maintenance and care for exterior building materials was a key in the design and it affected the overall landscape design. To achieve the highest quality for the exterior spaces, strategies including superior amenity areas, spaces for renewable energy, stormwater absorption and control through the use of green roofs, and producing public and private gardens were implemented. Design aspects included flexibility, low maintenance, and a high level of privacy for the living and exterior spaces. The public gardens at the front of the site served as an extension of the existing front gardens that encompass the terrace of the houses in the southern part. A fruit tree planted in this garden surrounded by herbs and produce is seen as a strong visual stimulant from the street, thereby creating an image of an outdoor kitchen or even a market.

比弗兵营社区住房

The Beaver Barracks Community Housing

Ottawa, Ontario, Canada

Hobin Architecture Incorporated

The three catalysts for The Beaver Barracks Community Housing were sustainability, affordability, and accessibility. To meet standards of sustainable design, the initiators considered the social, environmental, and economic factors. This process involved careful planning from the concept and design phase to the building procedures to create an exciting communal landmark.

The goal of the client, Centretown Citizens Ottawa Corporation (CCOC), was to create a community with affordable housing that appealed to low- and moderate-income inhabitants. The development is located in the center of Ottawa, within walking distance from local amenities such as stores, schools, parks, and public transit. Built on an empty brownfield site with a total area of 4.30 acres (1.73 hectares), the Beaver Barracks Community offers 254 units that cater to singles, couples, families, and seniors. This promotes communal mingling of different generations in a vibrant atmosphere. Furthermore, 25 of these units are wheelchair accessible and an underground geothermal system is used for heating and cooling each individual living space. The community consumes 40 percent less energy than the average energy expenditure in a similar project. To attain this, a unique geothermal system, high-performance envelopes, thermally broken balcony slabs, high-efficiency lighting and appliances, and green-roof technology were implemented. In addition, the units utilize natural light, which is made easy through the open-interior concept design.

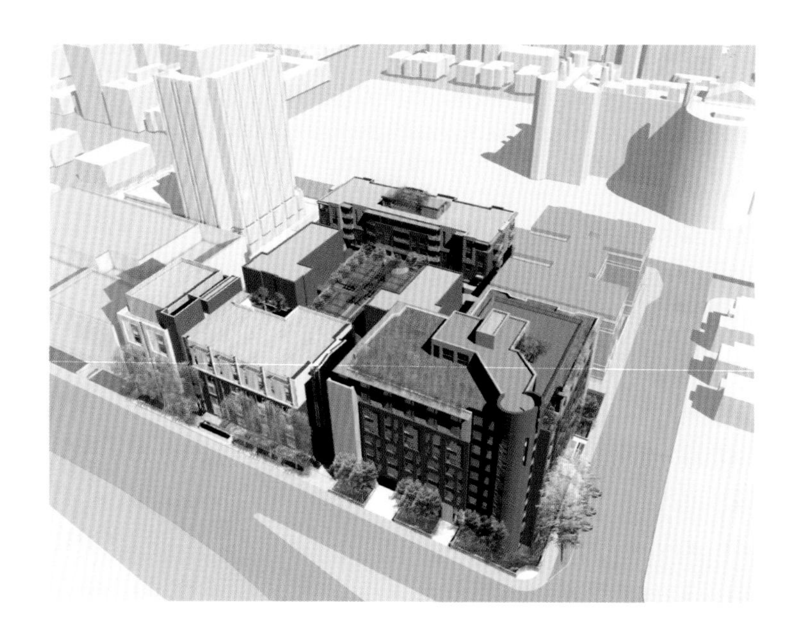

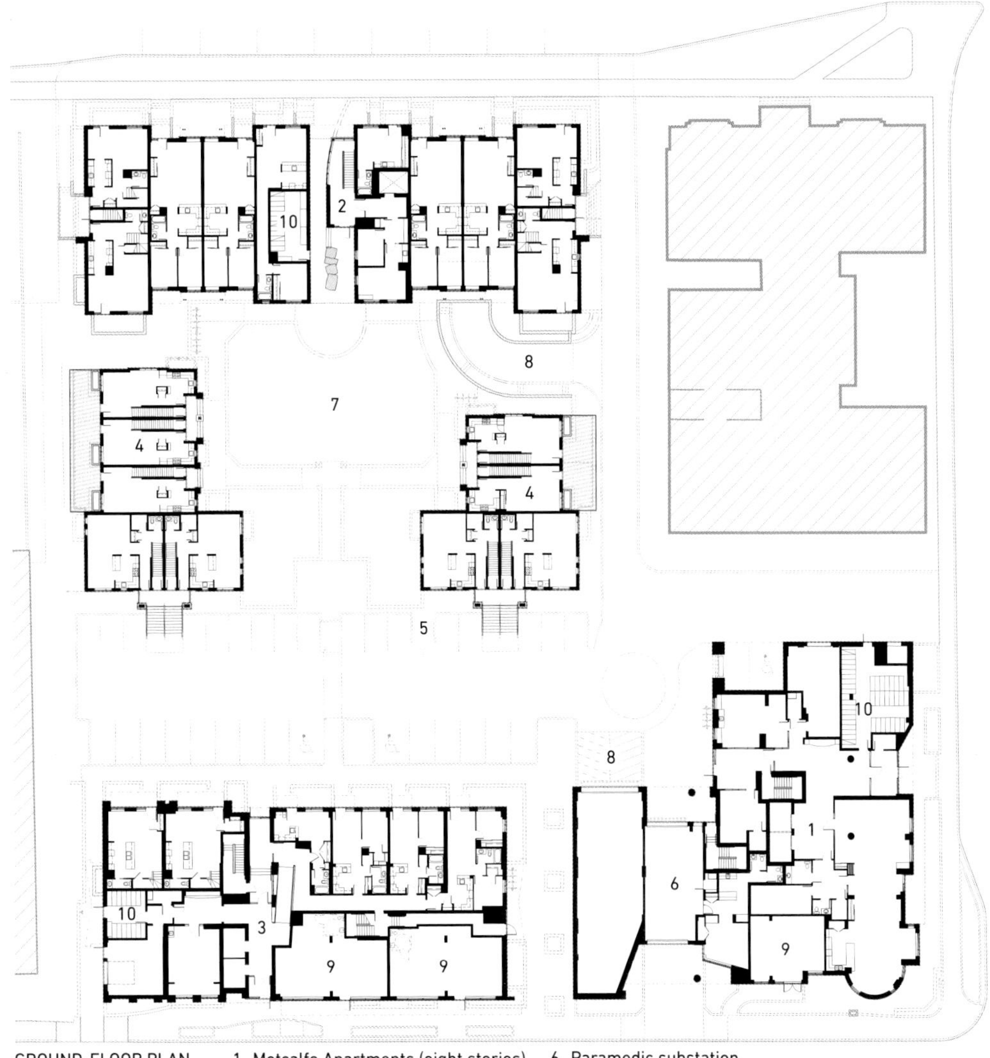

GROUND-FLOOR PLAN

1 Metcalfe Apartments (eight stories)
2 Argyle Apartments (four stories)
3 Catherine mixed-use (seven stories)
4 Stacked townhouses (four stories)
5 Geothermal plant (underground)
6 Paramedic substation
7 Community garden
8 Ramp to underground parking
9 Leasable retail space
10 Secure bicycle storage

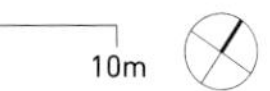

As previously discussed, sustainable developments that reduce the harmful effects of climate change together with the residential care of the environment allow the community to flourish. The inhabitants sign "green commitments" to pledge their support for "green activism" within Beaver Barracks and even beyond. To promote this "green activism" residents can easily obtain free compost buckets, get energy savings tips, free bike storage, premium VrtuCar parking, and a focal point garden program, which the buildings have been arranged around. Furthermore, alternative forms of transportation such as public transit, cycling, and walking are encouraged.

An open courtyard that encompasses the center of the community provides an exterior gathering place for residents rich with natural beauty and open space. Plants in this space need minimal maintenance and the exclusive community garden encourages inhabitants to grow their own produce. A second, small community garden and green roof sit on top of the tallest building, thus minimizing lost space and providing another opportunity for residents to socialize. One resident states that the community and their personal vegetable gardens are a great contribution to the neighborhood. Beaver Barracks can be defined as a model for sustainable site planning and community development.

6
9
10
7

102
315 Victory Gardens Private
101

VICTORY GARDEN

占地社区

Landgrab

Shenzhen, China

Jose Esparza, Joseph Grima, Jeffrey Johnson

Landgrab was built during the 2009 Shenzhen Biennale of Urbanism and Architecture under the title "mobilization." The heart of downtown Shenzhen was transformed into a square urban farm representing a small map of the city. Joseph Grima, Jeffrey Johnson, and José Esparza from SLAB envisioned the project and sought to represent the amount of land required to feed Shenzhen's 4.5 million people. The small map represents the city area and its borders while the farms represent, at the same scale, the total agricultural land needed to feed this same population. The farm is also subdivided into the various food groups: vegetables, fruit, cereals, etc.

The Landgrab experiment rethinks the use of urban farming. While feeding a small part of the local residents, the farm also raises awareness of food shortage issues on a larger scale. More importantly, its shock value comes from its prominent location and its spatial definition of the 21st-century sustainable issues.

It can be said that Landgrab serves a less dramatic purpose. It demonstrates that urban farming can be an aesthetic alternative to the traditional city square.

Indeed, just as Landgrab was intended to be a showcase, city farms can be as interesting and confrontational as public art. In other words, while rooftop or underground farms tend to be hidden from sight, a prominently located farm can help spur the awakening of the public's consciousness.

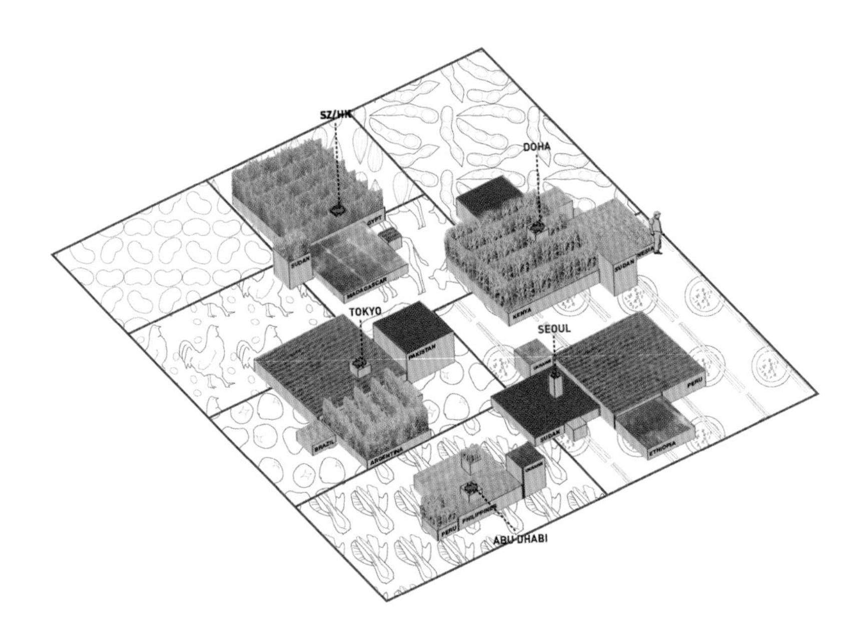

艾尚健康
妍美茉莉
年輪
Buy

滨海眼镜商场
北大荒绿色食品

妍美茉莉
养身美容会所
养生馆

美容连锁机构
滨海眼镜
滨海眼镜商场

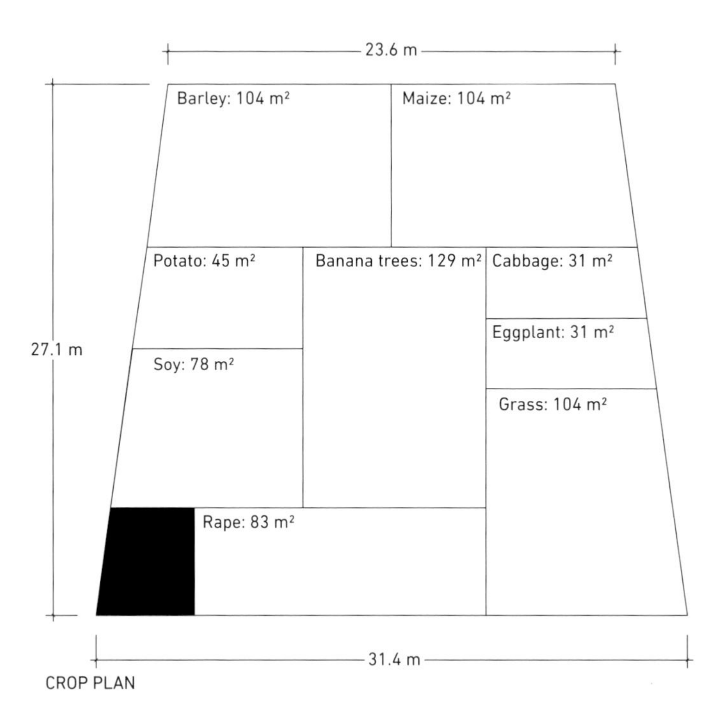
23.6 m
Barley: 104 m²
Maize: 104 m²
Potato: 45 m²
Banana trees: 129 m²
Cabbage: 31 m²
Eggplant: 31 m²
Soy: 78 m²
Grass: 104 m²
Rape: 83 m²
27.1 m
31.4 m

CROP PLAN

宽街18号社区
18Broadway

Kansas City, Kansas, United States
HOK

Designed by HOK, the project 18Broadway is located in the historic Crossroads Art District of downtown Kansas City. It is a corporate-sponsored urban rain garden that became the jewel of its community. The square-block garden not only helps the city with stormwater management, but also helps to showcase food growing techniques and renewable-energy technologies to the general public. It is where the water runs clear, the food is abundant, the shelter resourceful, and the energy endless.

Four sections of the garden are categorized and separated by function covering water, food, shelter, and energy. For water treatment, features include a storm planter, rain garden, and alley swales. For food, there are the demonstration and production gardens. A series of sustainable housing initiatives including parking stalls covered by green roofs compose the shelter section, as for energy, there are multitudes of alternative energy charging stations from compressed natural gas, to biodiesel, to electric.

A total of a million gallons of rainwater falls onto the site every year, a formidable quantity that, if harvested correctly, could benefit the neighborhood significantly. Initially 18Broadway was designed to emulate Kansas City's 10,000 rain gardens initiative to help the city's storm and wastewater treatment problems, but it has become much more to its local community. It not only uses rain gardens, shallow filtration basins planted with deep rooted vegetation that catch and filter pollutants from roadway and roof runoffs, allowing the water to be slowly absorbed by soil, but also uses an underground ultraviolet (UV) purification system so that the water when reused is clean and pure for the plants that, in turn, will bear fruit.

The project was an attempt to solve an immediate problem in the urban environment, but has additionally provided more opportunities and potential for a green and alternative space where positive values and methods are showcased and put into action. 18Broadway in Kansas City not only helped to demonstrate how to use rainwater for food production, but also how urban agriculture works as well.

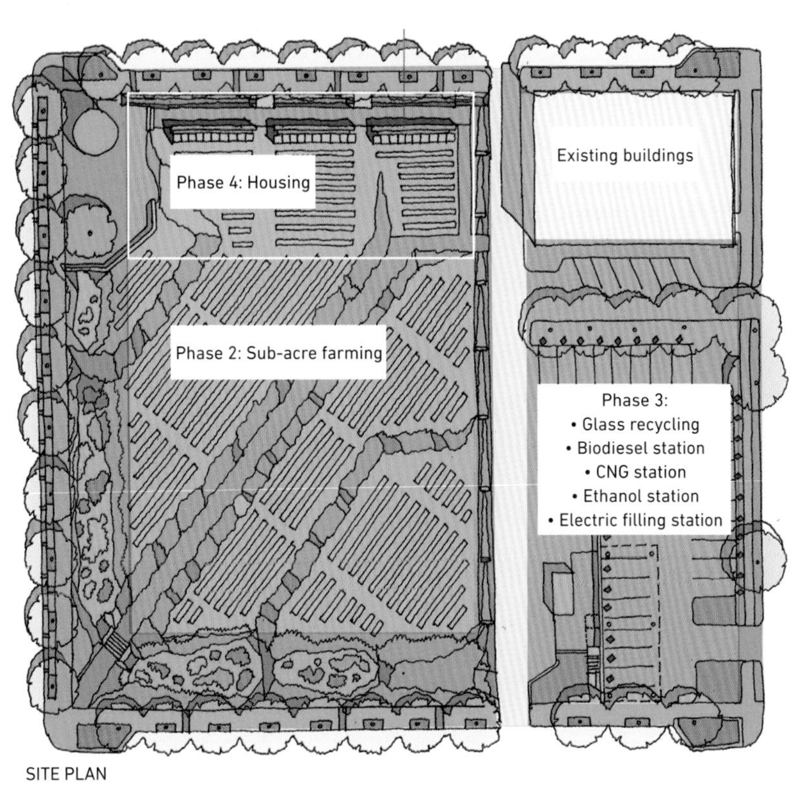
Phase 4: Housing
Existing buildings
Phase 2: Sub-acre farming
Phase 3:
• Glass recycling
• Biodiesel station
• CNG station
• Ethanol station
• Electric filling station

SITE PLAN

第八章：区域供热

DISTRICT HEATING

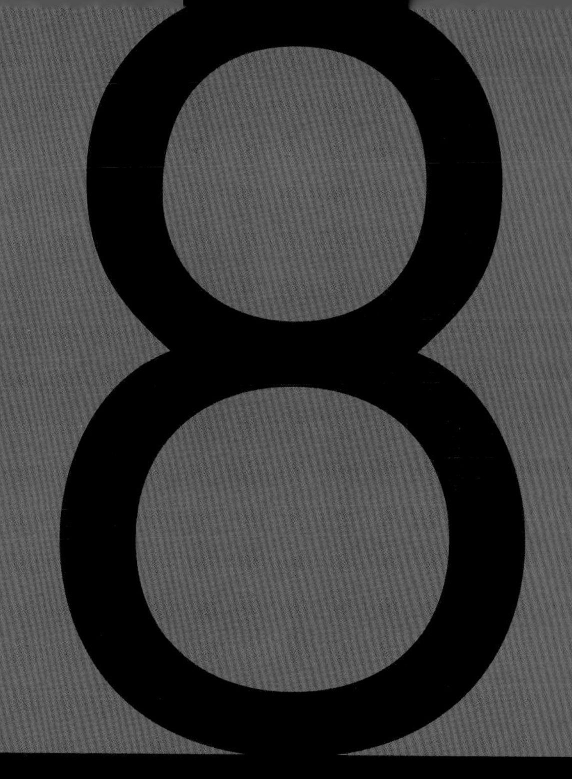

The need to rethink the use of non-renewable energy sources is becoming increasingly urgent due to their depletion and the challenge global warming poses. Also, contemporary methods of heating and cooling for individual dwelling units cannot be seen as efficient or environmentally friendly and their viability must be seriously questioned. District heating systems, on the other hand, supply multiple homes from a single central source making them highly efficient. This chapter examines the advantages and disadvantages of these systems, outlines their principles, and examines different source and distribution types.

District heating is the generation and distribution of thermal energy at the community level (Enwave 2011b). Entire neighborhoods are supplied by a single external source and are connected to a network rather than each building having its own energy source. A central plant is where the heating, and at times the cooling, for an entire neighborhood is generated. A district heating plant produces energy with relatively minimal loss, which is then distributed to individual homes via a network such as underground pipes. Any leftover energy is routed back to the central plant to be stored for later redistribution as needed.

Options for simultaneously generating heat and power are also common. The main purpose of district heating, however, is to produce energy for heating and hot water. A typical district heating scheme can be divided into three parts: production, distribution, and consumption (Dalkia 2009). When a district heating system is properly planned, it will be easy to run and maintain (Euroheat 2011). Moreover, homes that are connected to a district heating plant usually have a meter that measures consumption encouraging people to be more energy-efficient (Burke 2012).

District heating systems are low-maintenance and have a long lifespan, an advantage in neighborhoods. The property values of homes connected to district heating will consequently increase and space that would have been used for mechanical systems in each unit will be freed (Enwave 2011b). Costs are generally lower in district heating than traditional methods. Indeed, when natural resources such as sun and wind power are used as input energy sources, there is no cost associated with sourcing. Another advantage of district heating is the ability to use renewable resources and thus produce zero greenhouse gas emissions. Air quality in the neighborhood is improved and air pollution, if any, is reduced. These systems also have minimal heat losses, further contributing to their efficiency and sustainability.

Energy security in neighborhoods also improves with district heating, especially in countries that import energy (Euroheat 2011). District heating systems are highly reliable because they use a wide range of sources, leading to a lower risk of failure in delivery. For example, in 2008 the average heat-delivery break-time per household in Finland was 1 hour and 45 minutes, for an accuracy of 99.98 percent (Euroheat 2011).

Despite the many advantages of district heating, there are some disadvantages. First, a considerable initial investment is required to build a district heating plant. An already established neighborhood with busy roads will experience traffic disruption if a new district heating plant is to be constructed, not to mention that it can be costly and inconvenient. Pedestrians and drivers may have to take detours during their daily activities. Furthermore, households must be willing to change their heating systems to district heating, which might call for an upgrade in household insulation with its attendant costs, for example.

Heat losses must also be considered, however minimal they may be. Although breaks are rare, any failure in a district heating system will result in service interruption for the whole neighborhood. Lastly, a neighborhood must have a utility service to run the district heating system that will commit to a long-term engagement and service (Burke 2012).

In terms of sources for district heating, a common one is surplus heat. Processes such as the generation of electricity or the burning of waste will release heat as an energy surplus. District heating systems recycle this energy instead of letting it go to waste. The surplus heat is recycled and distributed to each home, which can then be heated. Other kinds of renewable-energy sources that can be used in district heating include wind, biomass, geothermal, and solar (Euroheat 2013). Moreover, district cooling systems will similarly make use of local resources such as the natural cooling from seas, lakes, and rivers. The main idea of district energy systems is, therefore, to collect and channel local energy such as leftover heat, renewable energies, or fossil fuels. In sum, these systems are inherently sustainable, creating communities that require less energy input and generate fewer emissions.

District heating systems may take energy from many different types of sources and redistribute them as appropriate. Underground pipes are a primary method by which energy may be distributed to buildings in a neighborhood. An elaborate network is created to connect the central source to individual houses. Hot water, steam, or cold water may then travel through the pipes. The water is heated or cooled using a specific energy source at a centralized location and is then distributed to connected buildings within the neighborhood's network. The water will thereafter return to the main plant to be reheated. The pipes are often insulated and installed in trenches connected to each home (Harvey 2006).

Leftover energy from the generation of electricity is a useful source for district heating systems. For example, neighborhoods can use leftover exhaust heat from nearby industrial processes. For instance, burning fuel to produce heat will give rise to two outcomes: either electricity or else heat that is dissipated to the surrounding. The energy that is dissipated to the environment is what can be used in district heating systems. Exhaust gas is another type of leftover energy that can be recaptured for use in district heating systems. Neighborhoods can, as a result, be heated by a centralized system that uses leftover energy released during different electricity generation processes.

Large bodies of water found in nature also make useful energy sources. They are especially relevant when discussing district cooling systems. The surface of a lake, for instance, will cool to 39.2°F (4°C) in the winter. Cold water is denser than warm water. Therefore, in winter surface water will sink to the bottom of a body of water (Enwave 2011a). Conversely, surface water in the summer will remain at the top because hot water is not as dense as cold water. A lake is, therefore, a natural reservoir of cold water and is a renewable resource. It is the coldness of the water that is harvested, rather than the water itself (Enwave 2011a). A neighborhood that makes use of a nearby lake as a source for district energy systems is contributing to the sustainable use of natural resources by reducing the demand for individual air-conditioning units.

Geothermal energy is another energy source for district heating systems. The earth is comprised of different layers: the core, outer core, mantle, and crust. The core is made of iron, the outer core of magma, the mantle of magma and rock, and the crust by the earth's oceans and continents. The thickness of the crust varies, since the crust is made of separate tectonic plates. At the earth's core, geothermal heat is produced, generated by the slow decay of radioactive chemical elements that make-up the earth (Mannvit 2013). This energy can be used in district heating systems because of the movement of the earth's tectonic plates. These plates will slide, collide, or pull apart, and magma will be brought to the surface. As a result, geothermal heat will be rendered accessible to neighborhoods for energy harnessing. Geothermal energy can be found anywhere, and is a reliable source for district heating systems. Neighborhoods that take advantage of geothermal heat contribute to preserving the ecosystem, since making use of the environment's natural resources will help reduce dependency on fossil fuels.

Solar energy can be harvested as a source for district heating systems as well. In principle, solar heat is available anywhere, although depending on the region some neighborhoods might have more access to sunlight than others. Solar collectors are mounted on a roof or on the ground. These collectors are then connected to a circuit that circulates the energy to households in a neighborhood (CIT 2010). It is important to note, however, that solar energy is less available during the winter. Hence, to increase the potential use of solar energy in district heating systems, it is necessary to store the solar heat harvested during the summer. Collector arrays and storage are often built in close proximity to one another. Solar district heating plants in neighborhoods are sustainable because they require little maintenance after construction, and they produce zero greenhouse gas (GHG) emissions because they make use of the sun's heat.

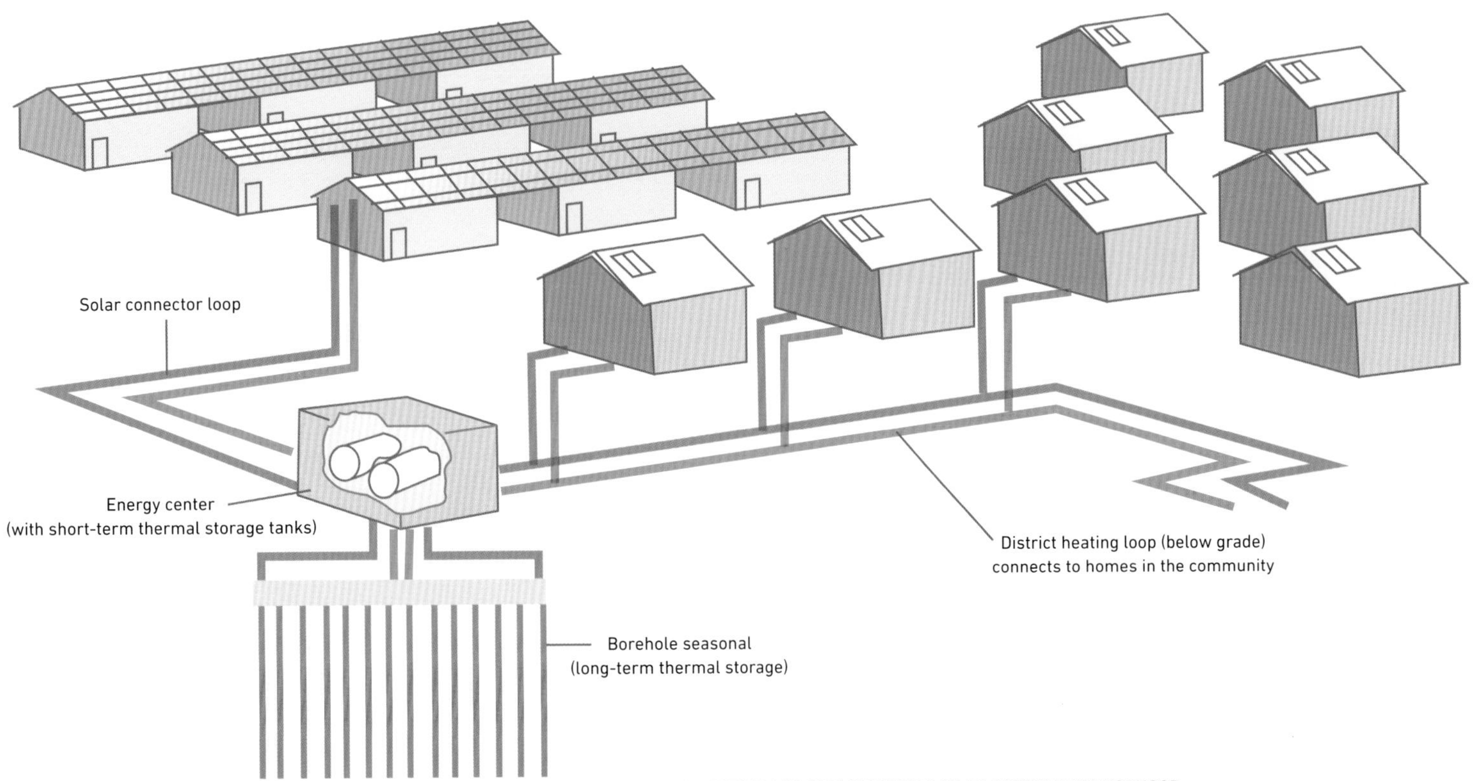

DISTRICT HEATING NETWORK IN A COMMUNITY WHERE A CENTRAL PLANT GENERATES HEATING AND COOLING NEEDS OF AN ENTIRE NEIGHBORHOOD.

Wind turbines can power an entire neighborhood by converting wind into energy. Again, some neighborhoods have higher wind speeds than others and there are also times where there is little to no wind. Energy produced by wind, however, can be stored for several days (Nordic 2010). In principle, wind energy is first produced by the turbines which then will be used at the central plant to heat water. The turbines can be directly connected to heat pumps, and the energy distributed through the network (EA 2011).

District heating systems are an innovative way to deal with issues related to global warming. Neighborhoods with district heating systems help reduce primary energy demand and contribute to the sustainable use of natural resources. Change, however, may be slow since district heating requires a considerable initial investment.

朱比利码头社区

Jubilee Wharf

Penryn, United Kingdom

ZED Factory

Jubilee Wharf is an award winning development by ZED Factory. It is located in the heart of Penryn, a small city in Cornwall, beside the Penryn River in the United Kingdom. The site was a derelict former coal yard, which was transformed into a small mixed-use development. The main goal of the designers was to create a focal point to foster a sense of community through a range of activities.

Jubilee Wharf consists of two buildings by the water, a taller one behind a shorter one so that both may enjoy views of the river. They are separated by an intimate courtyard in between. The taller building houses 12 workplace units that are rented out to local artists and craftsman and six residential duplexes. The shorter building houses functions such as a café, a community hall for events and fitness lessons, and a nursery. The development was designed to be socially inclusive.

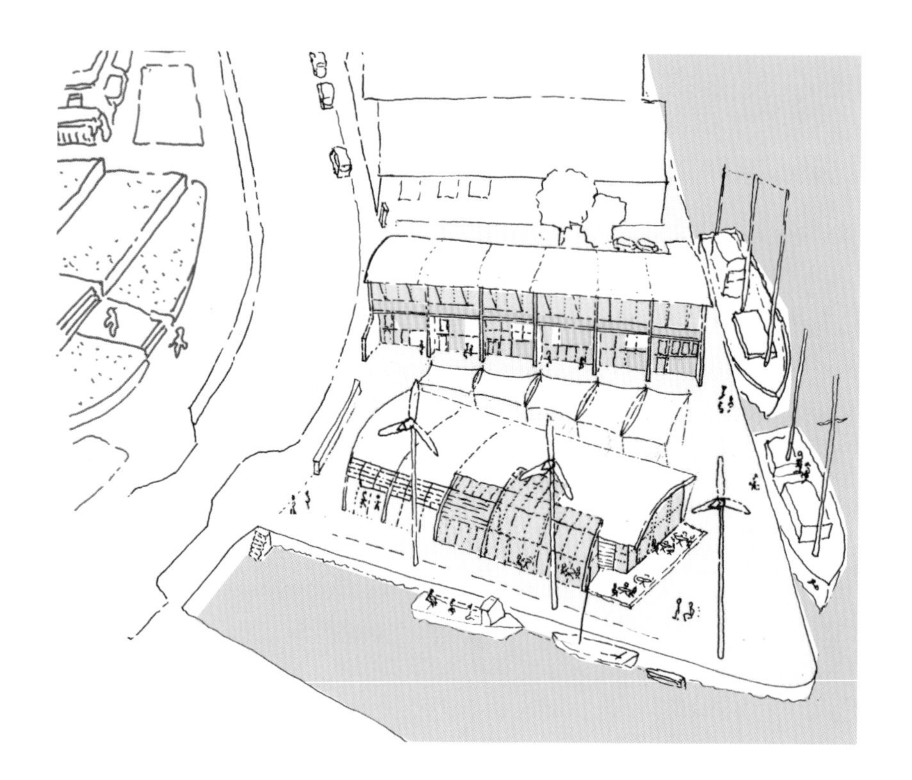

Jubilee Wharf
JUMBLIES
Truro
B3292
Falmouth

JUMBLIES

Jubilee Wharf

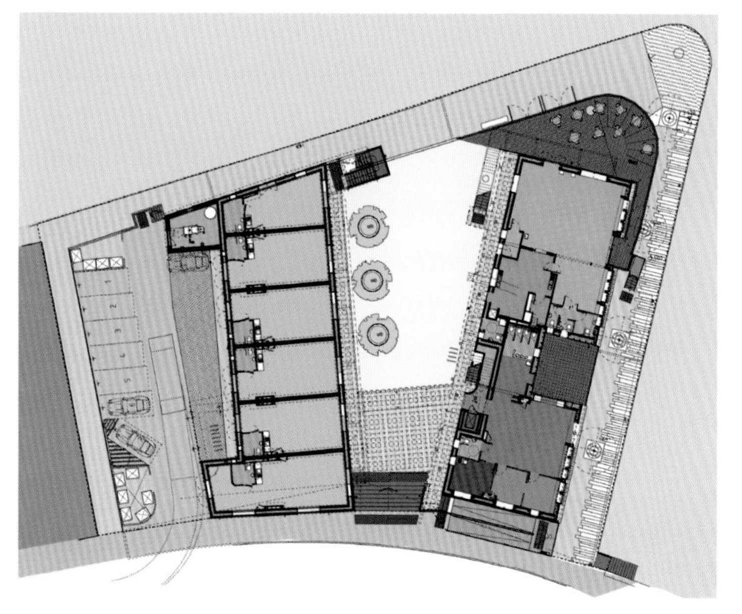

GROUND-FLOOR PLAN

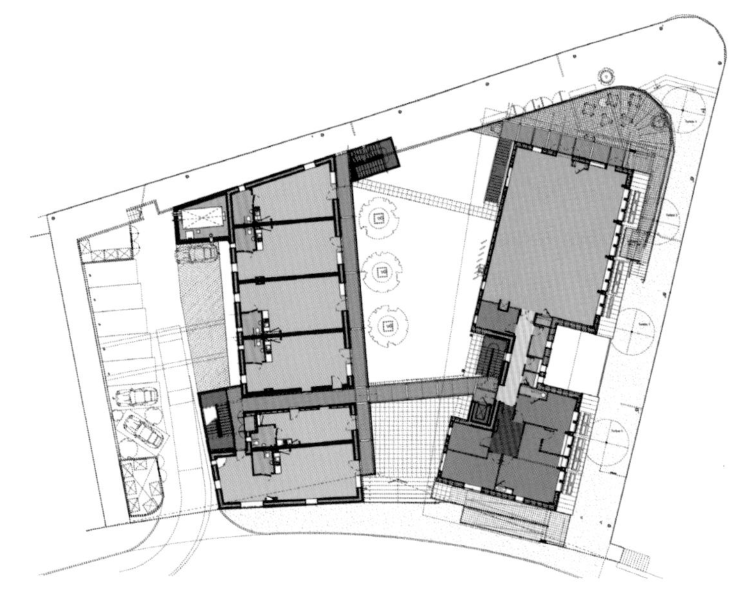

FIRST-FLOOR PLAN

The plan of the project was to be environmentally responsible with near zero carbon emissions, and appliances with low energy consumption. Four wind turbines generate electricity to take advantage of the strong winds on the river banks. A small wood-pellet boiler is used for underfloor and domestic hot-water heating, and is supplemented by solar panels. Only local or reclaimed materials, such as old floor boards and granite, have been used. The two buildings are also well insulated helping to decrease energy consumption.

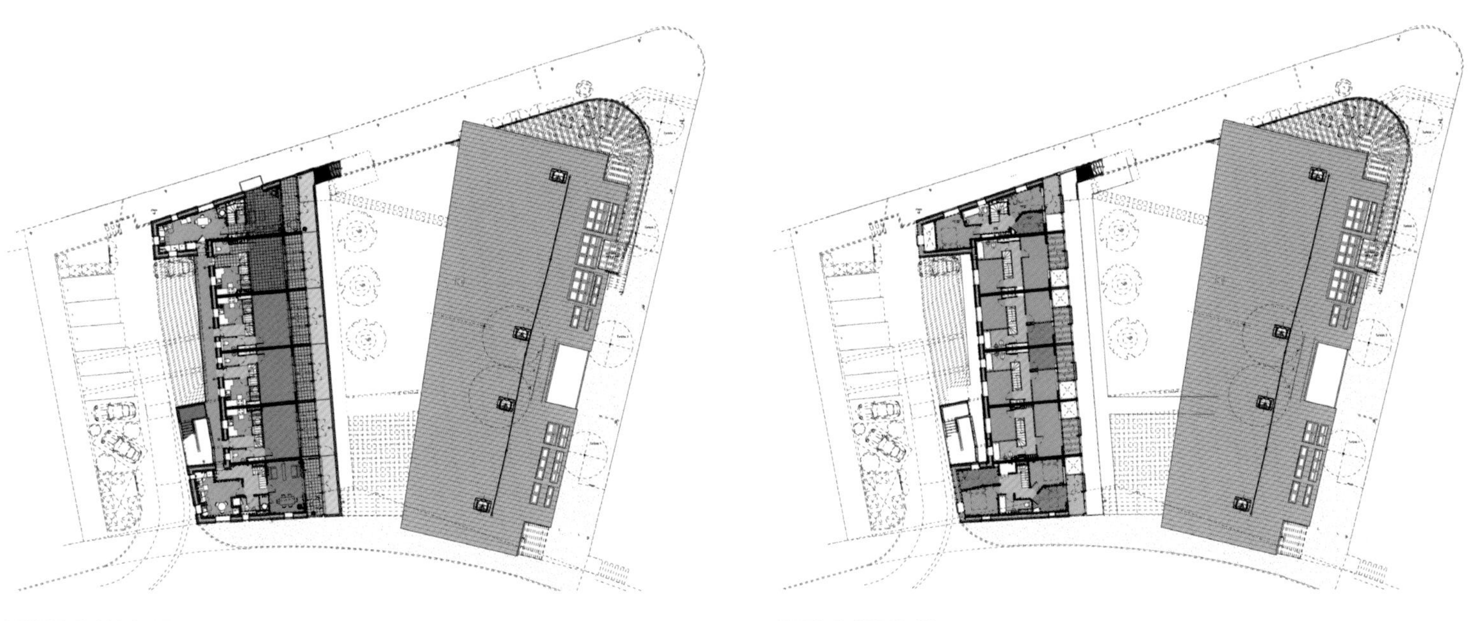

SECOND-FLOOR PLAN

THIRD-FLOOR PLAN

弗莱堡太阳能拓居地

Solar Settlement in Freiburg

Freiburg, Germany

Rolf Disch Solar Architecture

The idea for the Solar Settlement was conceived during Expo 2000 in Hanover, Germany, and was built six years later on the site of the former French Vauban barracks, 1.9 miles (3 kilometers) from Freiburg city center in Germany. The project consists of an eco-village of 90 units that relies solely on the solar energy gathered on a site of 103.78 acres (42 hectares). The eco-village was designed as a network of intimate streets and private gardens. This mixed-use urban neighborhood consists of 80 percent residential and 20 percent workplace.

The most innovative feature of Solar Settlement is that each unit is a PlusEnergy home, meaning it produces more electricity than it consumes. The community generates a total of 596.5 horsepower (445 kilowatts) per year from the photovoltaic panels that cover all the roofs and are oriented towards the south at the optimum angle to absorb the most of the winter sun. The surplus energy is then sold at the minimum price of 65 cents/kwh back to the grid, generating a source of additional income for the tenants. In total, about 53,000 gallons (200,000 liters) of fuel is saved with all the energy efficiency measures and the production of PV electricity.

Solar Settlement is a community in the best sense of the word. In addition to being the first plus-energy housing estate ever, the project also has progressive social programs. The complex is car free, with a communal parking garage, encouraging social interactions.

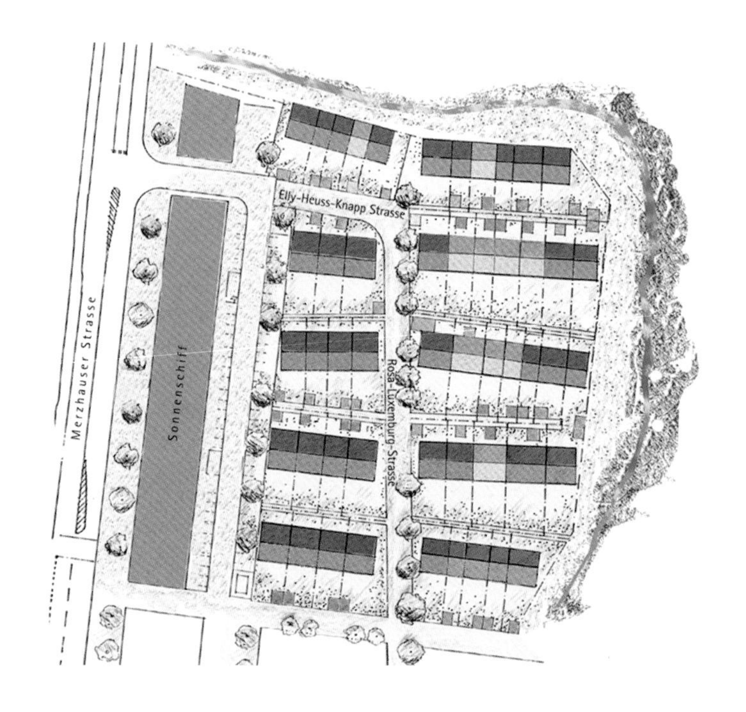

dm

There is also a car-sharing system that helps reduce the number of cars needed and establish connections between residents. Within the complex, there is a commercial building called the Sun Ship that houses a supermarket, pharmacy, communal underground garage, roof gardens, and penthouses.

There is a series of two- to three-story wooden row houses that are aligned to the south and built in accordance with the high requirements of passive house standards.

The Solar Settlement utilizes natural building materials and striking colors to add a distinctive look to the exterior. The implementation of energy-conscious construction, building waste management, cooperative mobility concepts, and sustainable water management work to give this example of solar architecture its environmentally friendly status and its social-minded design, helping make the neighborhood cognisant of basic human needs.

中央车站铁马社区

Ironhorse at Central Station

Oakland, California, United States

David Baker Architects, BRIDGE Housing

The Ironhorse at Central Station project was built in Oakland, California in 2010 as an affordable housing initative for people earning less than 50 percent of the area median income. It is part of the Central Station district, a historic area where Oakland's original train station was located. This section of Oakland is undergoing a large redevelopment that will see the transformation of 29 acres (11.7 hectares) of former industrial land into residential neighborhoods of over 1000 new homes, with Ironhorse including 99 units.

Comprising groups of four-story buildings that curve around community focal points such as the podium-level courtyard with a free-standing community pavilion, Ironhorse was designed to support "green" community living. Much of the previously existing urban plan had to be redesigned to include traffic calming pocket parks and pedestrian areas. Outdoor furniture, such as benches, has been made of recycled materials to include composite lumber.

FIRST-FLOOR PLAN

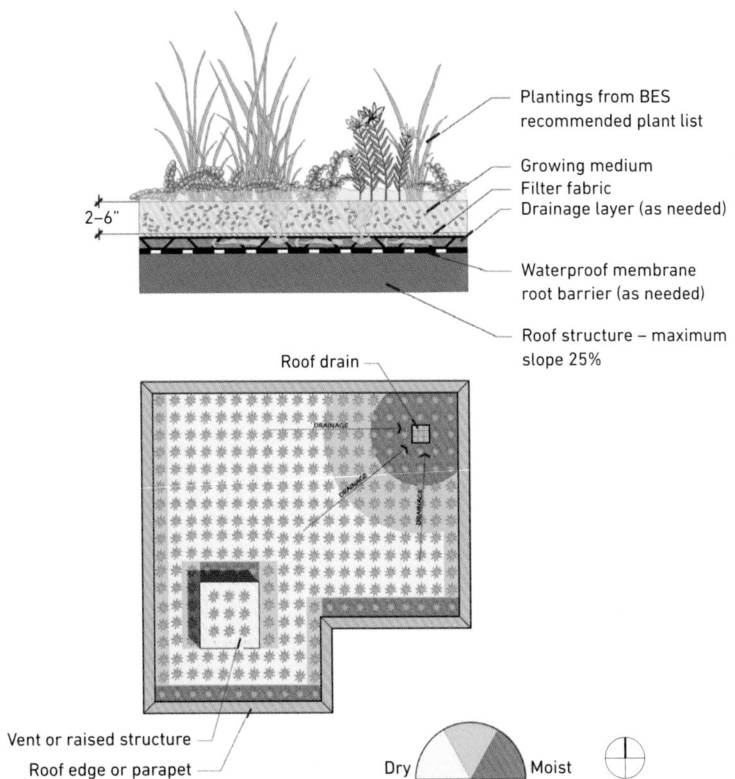

GREEN ROOF DIAGRAMS

The project includes highly innovative sustainable features. All electricity for the common areas is supplied by a 174-horsepower (130-kilowatt) photovoltaic system that will produce more than 301,730 horsepower (225,000 kilowatts) of electricity per year. Solar panels also provide about 60 percent of the heat for domestic hot water with the rest being supplied by energy from a high-efficiency gas-powered boiler. Some of the roofs are planted with vegetation that provides heat and sound insulation as well as providing places where residents can socialize. The landscape irrigation is dictated by the weather with control information coming in via satellite.

The Ironhorse could be said to be an exemplary project of high-quality housing that encourages a healthy living environment that reduces impact on air, water, and landfills. It transformed a previously desolate industrial landscape into a lively community.

奥斯特朗社区

Österäng

Kristianstad, Sweden

Sesam Arkitektkontor AB

Österäng, meaning "east meadow," is a large residential sector near the center of Kristianstad, Sweden. This district of 1200 apartment units was first built in the 1970s, but recently underwent a large-scale renovation. The final phases of the renovation were completed in 2010, transforming the complex into a sustainable community that supports a high quality of living.

The community comprises 28 three-story buildings and 18 eight-story apartments. Initially, there was much to change, from the overly monotonous and nondescript architecture to the poor quality cement, to sagging galleries and balconies, to insufficient drainage, damp bathrooms, unhealthy emissions of adhesives and plasters, and difficult access for the disabled. In addition to rectifying these problems, many new features were also included in the program. Recycling stations were added, paved surfaces were eliminated in favor of vegetation, filtering flora for rainwater purification was planted, and harmful building materials such as asbestos were removed.

For each unit, the roofs were transformed from flat to gabled and covered with zinc metal sheets and the façades were clad with panels of wood, fiber cement, or brick. Solar panels were added for hot water and two heat pumps were installed. The heating system was disconnected from the mains water and linked to the local biomass district heating network, monitored constantly by a computerized and centralized control system. Each individual unit came with its own heat recovery system and thermostatic valve to regulate heat and to measure energy consumption and ventilation.

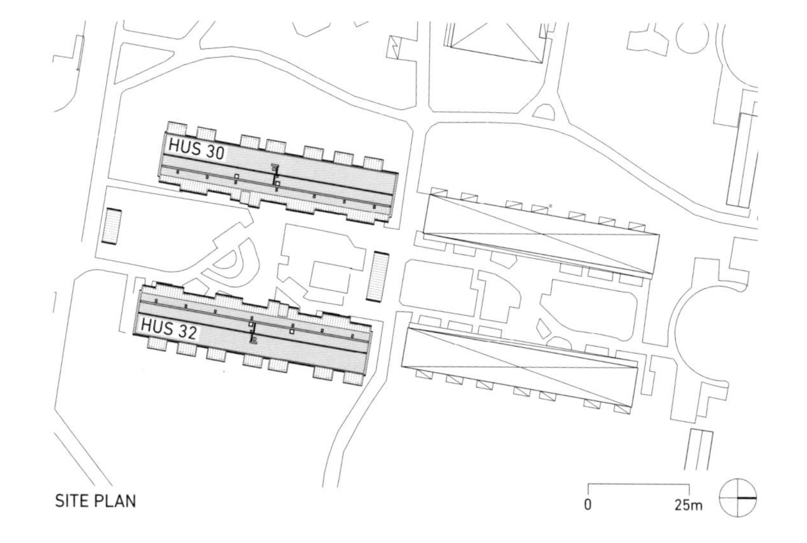

SITE PLAN

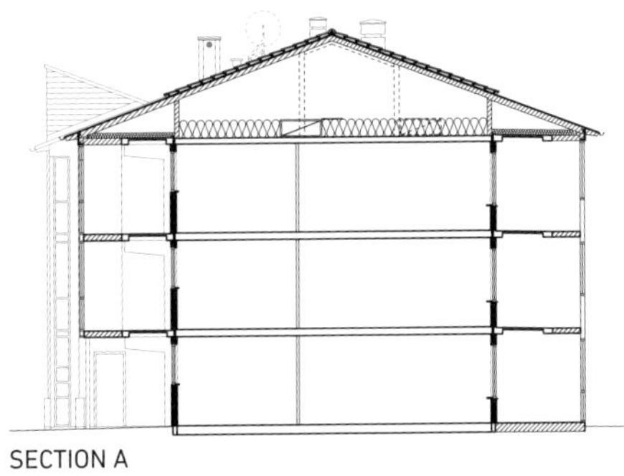

SECTION A

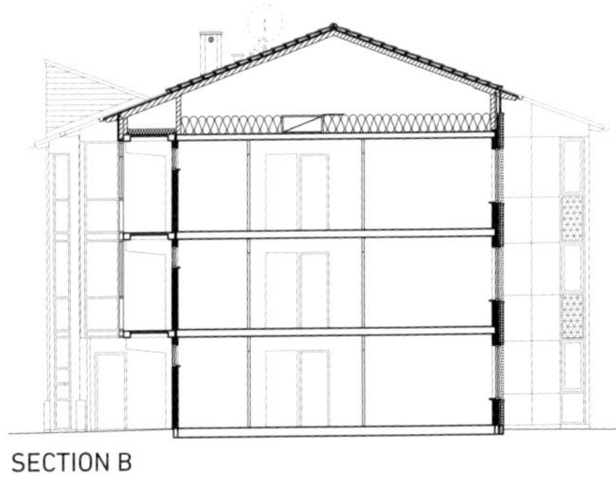

SECTION B

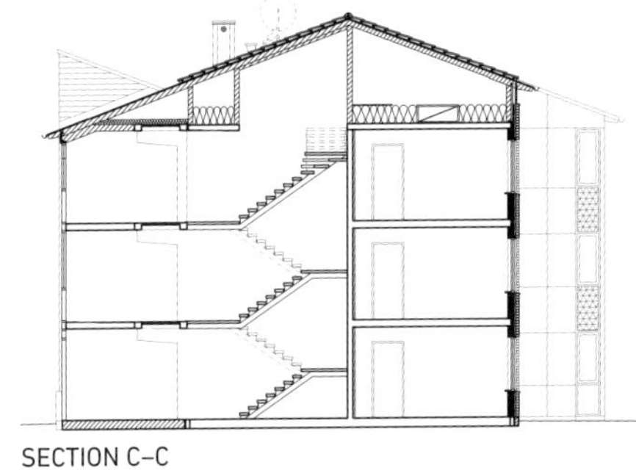

SECTION C–C

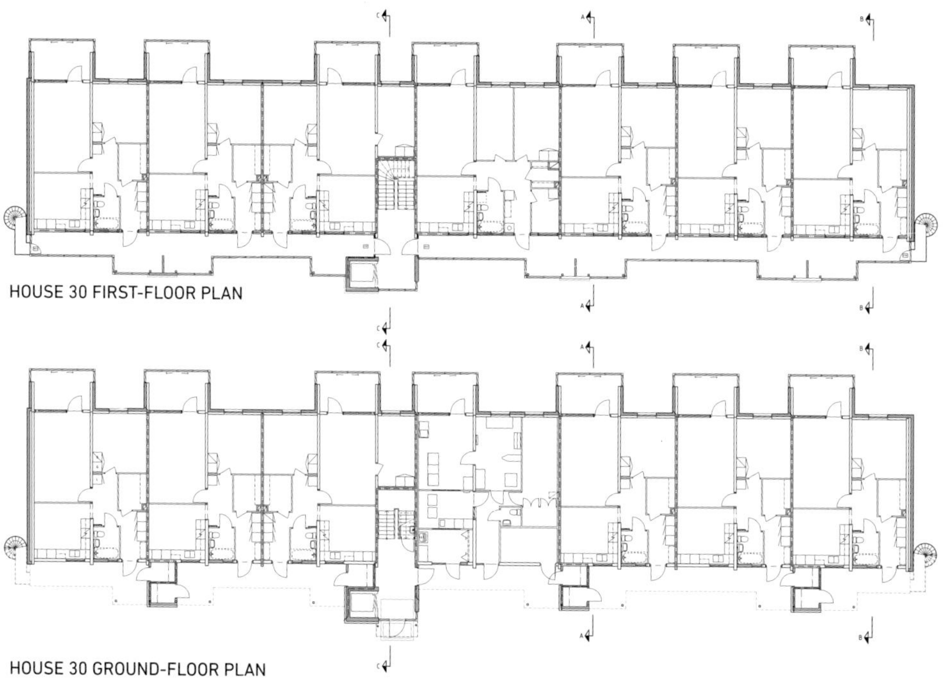

HOUSE 30 FIRST-FLOOR PLAN

HOUSE 30 GROUND-FLOOR PLAN

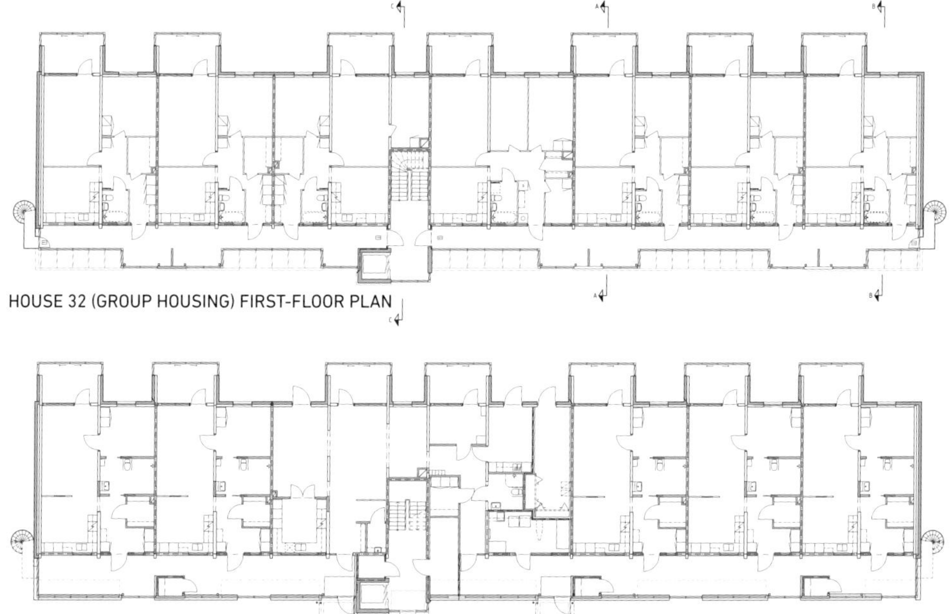

HOUSE 32 (GROUP HOUSING) FIRST-FLOOR PLAN

HOUSE 32 (GROUP HOUSING) GROUND-FLOOR PLAN

By integrating sustainable design principles in the renovation of the complex, a new level of integration was established between the residents and their environment. The building, through the recent work, extended its life by a century and eliminated much of the waste that would have been produced if the complex had just been rebuilt.

古什迪恩太阳能建筑

Solhusen Gårdsten

Göteborg, Sweden

Nordström Kelly Architects AB

The Solhusen (Sun houses) Gårdsten of Göteborg, Sweden, initially established in 1997, underwent some major renovations in 2004. The goal was to enable the tenants to be as self-reliant as possible in terms of the production and monitoring of the energy used in the 255-unit community comprising seven three-story buildings and three houses. The approach taken was to give as much say to the residents as possible, by working closely with the developers on technical, social, and economic analyses, and following an integrated design process.

Covering a total area of 4.7 acres (1.9 hectares), the renovation had brought on board new energy solutions with a 40 percent savings in energy consumption for heating and 30 percent for electricity as a result. It had a strong emphasis on energy conservation, ecological adaptation, and renewable energy. Among the steps taken were the inclusion of glazed balconies to reduce heat loss and increase floor area, installation of composting machines to convert household waste to soil, the installation of large common greenhouses along the exterior corridors of the south façade, and solar panels to heat domestic hot water, which would then be put into harvesting tanks in the basement.

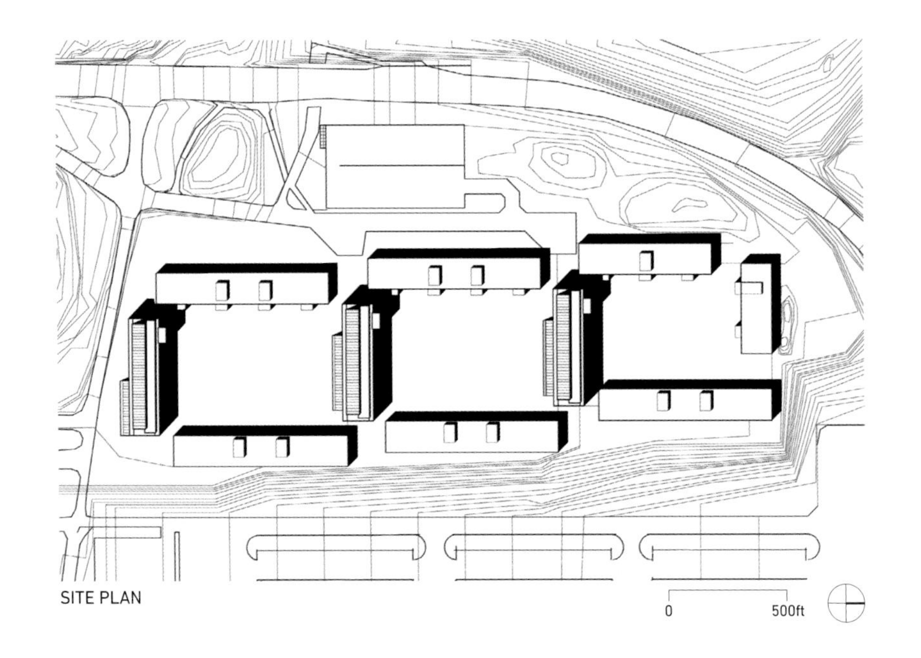

SITE PLAN

The application of modern recycling technology has also lowered the costs of housing and has led to improvement of environmental performance. The crowning achievement of the renovated units is that tenants can now control their own individual energy spending because metering equipment is provided for electricity, heating, and water. This increases user awareness, leading to a greater sense of responsibility and knowledge of energy uses and savings.

The project could be said to have attained its ambitious goal of reducing future maintenance and energy use. It has also garnered recognition and distinctions from the Swedish Association of Painting Contractors Prize for best-painted environment, the Great Energy Prize for individual metering, and the World Habitat Award of the Building and Social Housing Foundation.

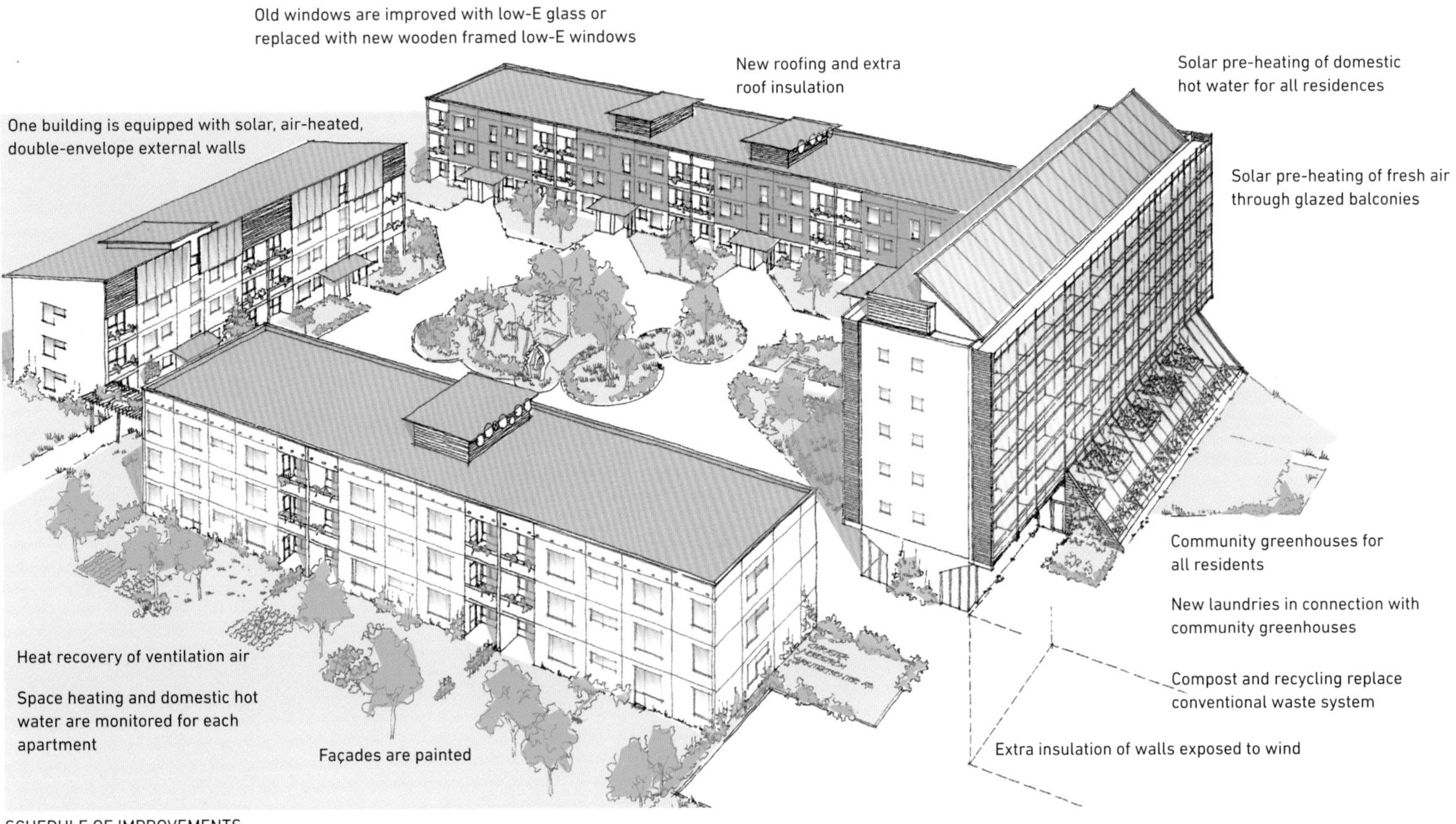

SCHEDULE OF IMPROVEMENTS

第九章：创新住宅概念

INNOVATIVE DWELLING CONCEPTS

Smart planning and design are often associated with systems such as photovoltaic panels, compost systems, and public transit networks. These aspects address environmental issues, but are only a partial solution to the challenges that society is facing. To be effective, sustainable development must also be extended to include social, economic, and cultural factors. This chapter focuses on the home and suggests changes that need to take place as a result of emerging contemporary trends.

In general, data demonstrates that the average size of a household in the western world has declined in recent years (Statistics Canada 2011). The combined number of households made up of seniors, singles, and single-parent families, which are known as non-traditional families, currently make up a larger percentage of the population. One of the effects of this demographic shift is the change in the dwelling types that these "new" population groups are seeking.

Homes built in North America since the Second World War have grown in area and occupied large lots (Friedman 2007). However, as a result of new economic realities in many nations, the number of homebuyers who are looking for lower cost units has risen sharply. Also, more people wish to reside near schools, parks, libraries, and stores to which they can walk. Unfortunately, such dwellings in those locations are not common. The need to advance sustainability by designing homes that are small, affordable, and energy efficient is necessary. A number of innovative dwelling concepts relevant to sustainable residential design and new lifestyles resulting from demographic shifts are described below.

Adaptable housing is sustainable because it responds to the changing needs of its occupants over time with minimal cost. It addresses current demographic shifts, where households are becoming smaller and older. The units are designed to facilitate accessibility by adjusting corridors, doorways, bathrooms, and kitchens to allow common occupants and people with disabilities to use them. The installation of grab bars in bathrooms is also considered, as well as electrical outlets, doors, and faucet handles that are positioned for easy access and use. People, as a result, are able to stay in the same house throughout their lifetime.

The benefits are not limited to people with disabilities. Entrances with ramps or with no steps create easier access for families with children in strollers, and wider doorways make it easier to move wide objects. Adaptable housing may, in addition, accommodate needs such as increased lighting levels for people with sensory disabilities. Windowsills may be lowered, as appropriate, to provide unobstructed views of the outside while sitting or lying down.

Adaptable housing is also cost-effective. For example, placing outlets and switches at reachable heights, and choosing lever-type door and sink handles involves little to no extra cost during construction. The cost of installing reinforcements during construction to allow for the eventual installation of grab bars in bathrooms is also relatively low.

Adaptable housing is furthermore sustainable, since energy and material use in remodelling is reduced. Also, because of the flexibility, it is an attractive option for a wide range of home buyers, and is a sound investment for resale and rental (Palmer 2010). Adaptable housing is not only limited to single-story houses; multilevel homes can still be suitable. Vertical lifts or staircase lifts to allow access to upper floors can be introduced as part of the design for adaptability process (District of Saanich 2012).

Another sustainable dwelling concept is multi-generational homes. These are homes in which three or more generations of the same family can reside. It is a way to not only address the demographic variations, but also to respond to new demands resulting from economic constraints. Indeed, the type of users might be couples with aging parents, an increasingly multiethnic population with a tradition of housing several generations under one roof, and even singles who may need to double up with siblings or friends in this fraught economic climate (Green 2012).

Social considerations for a multi-generational dwelling design include respecting privacy, maintaining independence, agreeing on childcare and discipline, and defining financial and household responsibilities (Atkinson 2013). Moreover, multi-generational dwellings contribute to sustainability since families living together will reduce the demand for housing. Services such as internet connections, heating, cooling, and water supply will also be shared. Generations of the same family will have a lower carbon footprint when living together than they would when residing separately. The time and car fuel that would have been expended for family visits is also reduced. Communities with multi-generational dwellings, therefore, offer collective sharing of resources.

Similarly, aging in place is another dwelling concept that involves a reduction in natural resource consumption. Dwellings are designed to support the needs of seniors, or to be adapted to do so. Seniors will therefore be able to continue living in their home as long as desired, because their home will provide secure and independent living. Housing design that provides for aging in place includes barrier-free design, handrails, and suitable lighting (Friedman 2012). An obstacle-free environment increases the ease of circulation within the home, and handrails that accompany stairs increase mobility. Moreover, ample amounts of lighting help maintain good vision. Aging in place removes the necessity to find new housing accommodations for seniors at retirement or nursing homes, which are often expensive.

By having dwellings that permit aging in place, resources need not to be spent on the construction of new retirement homes. Communities that encourage aging in place contribute to sustainability by giving a prolonged life to a home, which consequently reduces the necessity to build new homes adapted to seniors. Seniors are also able to maintain an independent lifestyle for as long as they desire, thereby increasing their quality of life.

Grow Homes are another dwelling concept that promote staying in the same place by allowing modifications to the size of the home. Not only can the positioning of interior partitions be modified, but the size of the dwelling can be expanded as well. Homeowners become part of the design process and have more say over the interior of their home. Grow Homes contribute to sustainability since there is a lower need for families to relocate to smaller or larger dwellings as the household composition changes. People can also continue to be part of and evolve in the same community. Consequently, consumption of energy and natural resources, a result of new construction, will be alleviated. Spaces in Grow Homes can be modified by using demountable partitions, or by attributing different functions to separate floor levels. For example, one family could reside on the first floor of a three-story building, while a second family could occupy the other two floors.

A unit can be expanded according to the add-in or add-on methods. When using the add-in method, certain space within the home—at first—is left unfinished, before being "grown" into later during occupancy. For example, an unfinished basement could eventually be converted into a family room. Second stories or attics are other locations for indoor growth. In addition to permitting homeowners to develop spaces gradually as needed, dwellings with unfinished spaces present a more affordable housing option. Indeed, after the cost of land, the second most expensive item of construction is the finishes (Friedman 2012). Materials are therefore saved and homes are only finished as needed by its occupants after the initial purchase. The add-on method anticipates and permits the eventual expansion of a dwelling beyond its original perimeter. Setback distances, maximum roof heights, andlot dimensions all define the expansion possibilities for a given unit.

Bushes and trees will need to be planted in places where they will not have to be uprooted later. The structure of the house must also be strong enough to support future additional loads due to the expansion of the home. Modular systems can be more easily disassembled than concrete, stone, or brick walls, and should be favored for add-on units. Grow Homes require rational design thinking, but successful initiatives may lead to many social, economic, and environmental benefits.

Live-work homes contribute to sustainability by allowing people to work from home. Instead of commuting to a workplace daily, the home takes on multi-uses. Live-work can provide good balance between career and family, since it offers the possibility for flexible schedules. To be efficient however, the home office must be designed to reduce distractions and promote productivity. The home, for example, should be separated from the workspace by partitions or different floor levels. Large windows that bring in natural daylight are effective in encouraging wellbeing and productivity.

Offices can easily be integrated into the design of any kind of home: a basement, the ground floor, or upper level rooms are some of the many options available to homeowners who want to work from home. In the past, working at home was only possible for a limited number of occupations, but recent developments in digital technology allow for a greater range of live-work opportunities. Moreover, live-work may contribute to reducing greenhouse gas emissions and therefore protecting the environment by eliminating the need to commute (Strickler 2012).

Small-sized houses also known as Micro Units also contribute to sustainability due to their sheer size. Compared with their larger-sized counterparts, small units require fewer

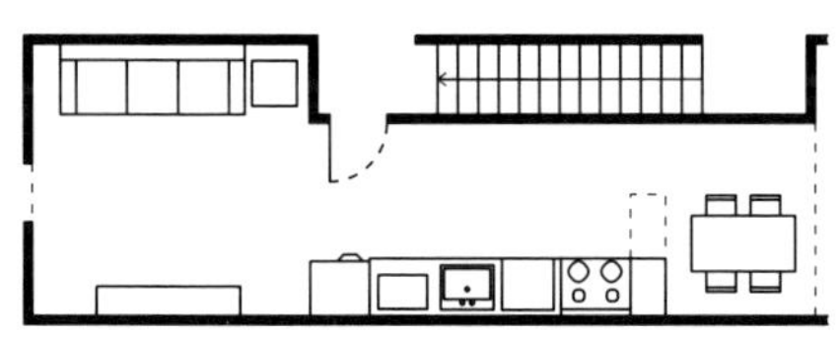
Open plan and minimum partitions allow a space to appear larger

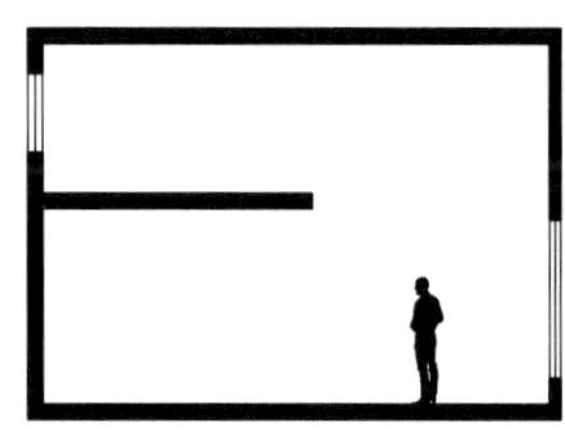
Changes in room height, length, or width delimit spaces without the need for partitions

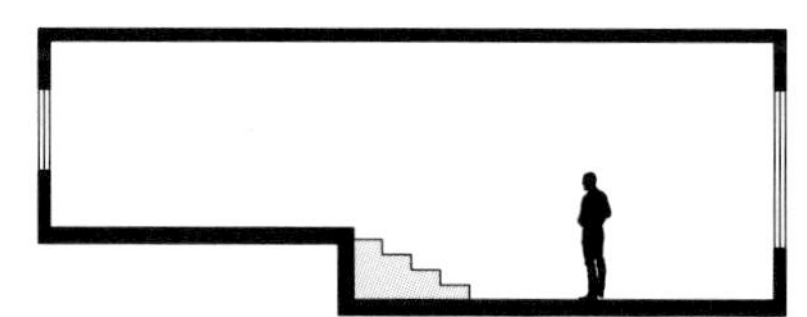
Steps and level changes underscore spatial and functional transitions

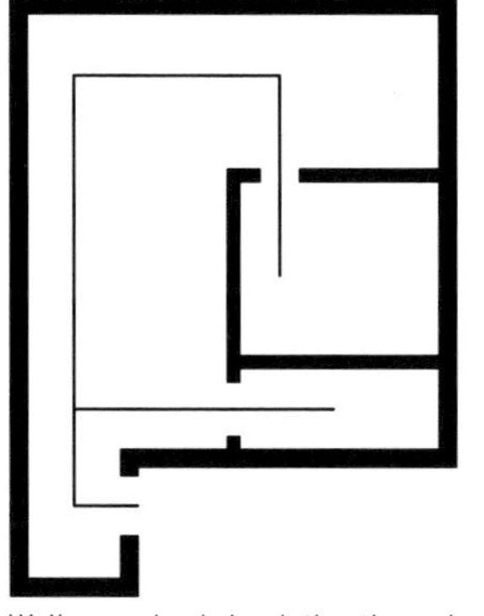
Well-conceived circulation through rooms eliminates the need for passageways

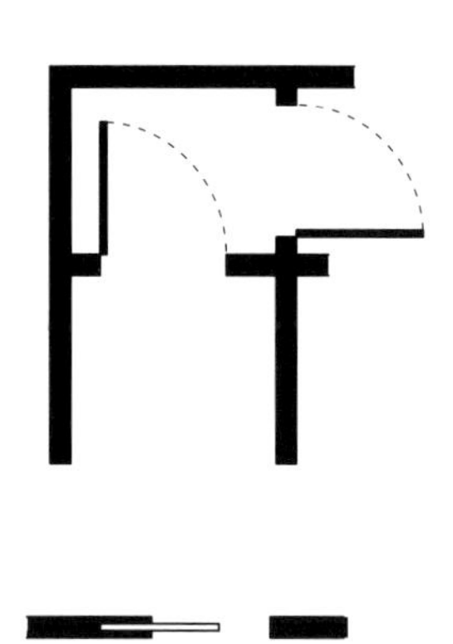
Pocket doors and doors that open outward facilitate and reduce circulation space

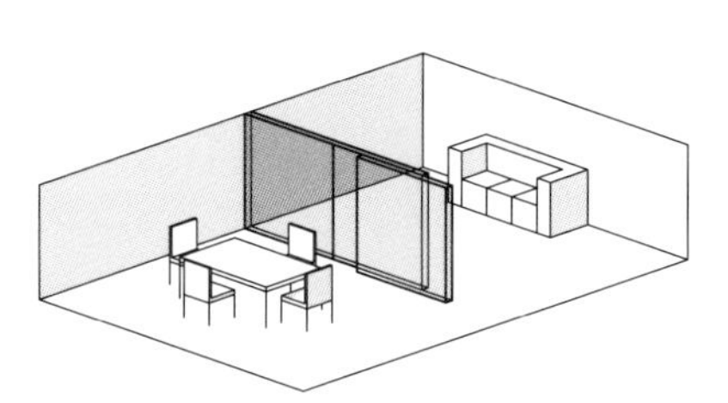
Sliding partitions or transparent materials make spaces appear larger

DESIGN STRATEGIES TO ENHANCE PERCEPTION OF SMALL SPACE IN A HOUSE.

materials for their construction, and consume less energy to heat and cool. Smaller houses occupy smaller plots of land as well, which reduces urban sprawl.

Neighborhoods with small-sized units help reduce resource and energy consumption. Small houses need to be efficiently designed to ensure the comfort and wellbeing of their occupants. Areas for basic uses such as sleeping, eating, and bathing need to be rationally planned. Large windows can blur the boundary between indoor and outdoor spaces, and higher ceilings give a more spacious feel. Large windows will allow in more natural sunlight, which in turn, is beneficial for the homeowner's wellbeing. In addition, space can be economized by minimizing interior wall partitions. Spaces can instead be defined by furniture placement, or by changes in floor materials or levels. Movable partitions are also an option if more privacy is needed. In sum, small-sized homes allow for flexibility in defining interior spaces even after construction. They are also a sustainable housing concept, since fewer resources and less energy are involved in their construction and maintenance than for larger houses.

Several dwelling concepts that present innovative approaches to smart sustainable living have been introduced above. These concepts reflect changes in housing demand due to social and demographic shifts. By addressing these social changes, houses are becoming smaller, more flexible, and are given multiple uses. Resource and energy consumption is in turn reduced, contributing to protecting the environment.

多尼布鲁克小区
Donnybrook Quarter

London, United Kingdom
Peter Barber Architects

The Donnybrook Quarter is located at the intersection of Parnell Road and Old Ford Road in East London. The pure white walls and wide lobby windows of the residences contrast sharply with the traditional brown stone façades and small windows of the dwellings across the street. Donnybrook Quarter was designed to alter traditional urban planning. The architect attempted to break up the homogenous town planning of London's terraced houses.

This project features 40 units of a mix of residential and live-work homes that range from one to three stories with a density of 45 units per 1 acre (0.40 hectares), attracting a variety of tenants. The first aspect one notices upon entering the site is its high density, even with the low-rise apartment units above. This allows for the maximum number of small-sized houses (i.e., micro units) to fit into a small area, contributing to sustainability by requiring fewer materials and consuming less energy to heat and cool.

The smaller one-story residences attracted seniors, while the three-story units, which combine live and work environments, attract artists or the self-employed and appeal to families with children. From the smallest to the largest, every unit has a terrace overlooking the streets.

The dwellings contain front windows with no transition at the building façades, in contrast to traditional English townhouses. This provides a smooth transition from private interior space to the public realm, which begins as soon as one goes out the front door. The units also contain small, private exterior balconies for enjoying the surroundings.

PARNELL ROAD E3

The architects recognized the importance of creating a space reserved for the community, even if it has meant that it is made-up of interior streets. The distribution of these streets, which helps organize the complex and makes walkability easier, also promote a thriving public space. All the balconies, bay windows, and roof terraces look out into these lanes, helping parents to keep an eye on playing children.

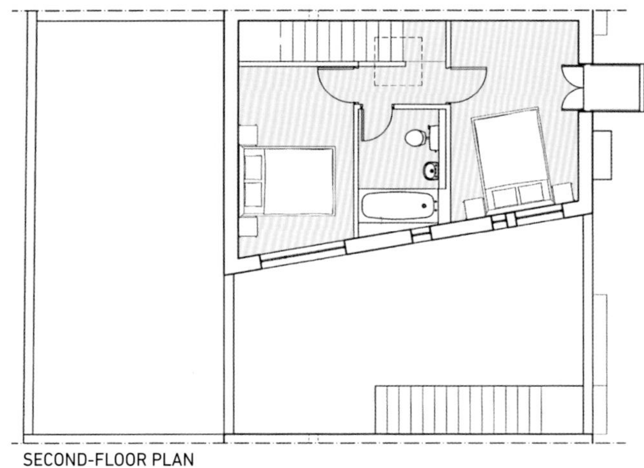

SECOND-FLOOR PLAN

SITE PLAN

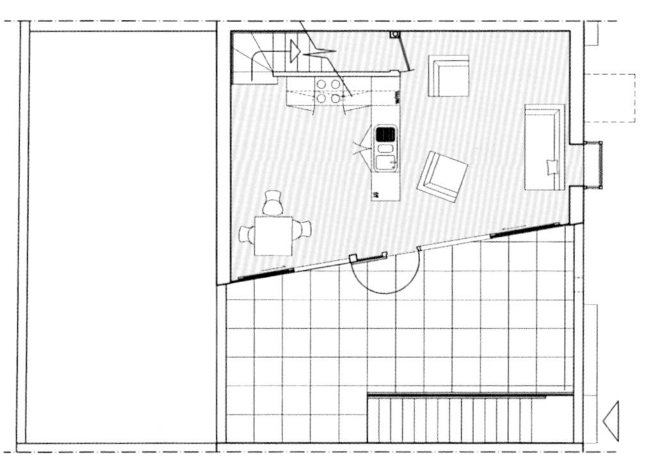

FIRST-FLOOR PLAN

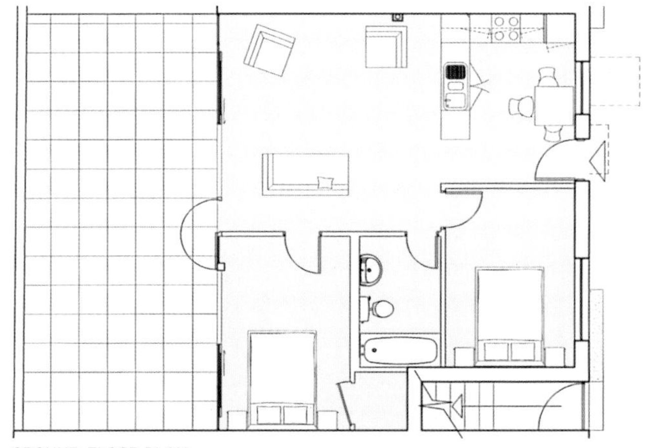

GROUND-FLOOR PLAN

郊区复兴生活区

Living Places Suburban Revival

Dandenong, Victoria, Australia

Bent Architecture

The Living Places Suburban Revival project was chosen from an open design competition initiated by the Office of the Victorian Government Architect and the Office of Housing. Living Places Suburban Revival was the culmination of an innovative approach to integrate medium-density public housing into low-density suburbs in an environmentally and socially sustainable way. The design set a high standard for sustainable low-cost housing in Victoria, Australia.

Comprising 15 dwellings that are spatially diverse, the project was designed with various floor plans. There was a need for flexibility due to the uncertain target demographics. As a result, the designers established a "flexible housing" concept which would respond to the needs of diverse changing populations over time.

Mediation between public open spaces, semi-private gardens, fences, and screens allow inhabitants to either connect with their neighbors or enjoy a private atmosphere within the unit, while experiencing outdoor communal spaces. Varied spatial arrangements create opportunities for interaction and individuality. These units have retractable wooden slat screens over windows and two-part fences where the top part remains open for neighborhood conversations. Existing vehicular and pedestrian paths were kept to maintain circulation rhythms. Many multipurpose areas were included in the complex to activate urban communal space and interrupt repetition. These outdoor areas come complete with barbecues, beautiful flora, and seating areas, which encourage social interaction.

To minimize the environmental impact of the project, methods such as greywater treatment, solar hot water systems, and photovoltaic panels for power generation and indigenous planting of vegetation were utilized. The project has received recognition from the Melbourne Design Award for Environmental and Social Credentials and others.

The Living Places Suburban Revival demonstrates that communal public housing can fit in with a suburban context and also accommodate individual and changing populations without hindering the neighborhood.

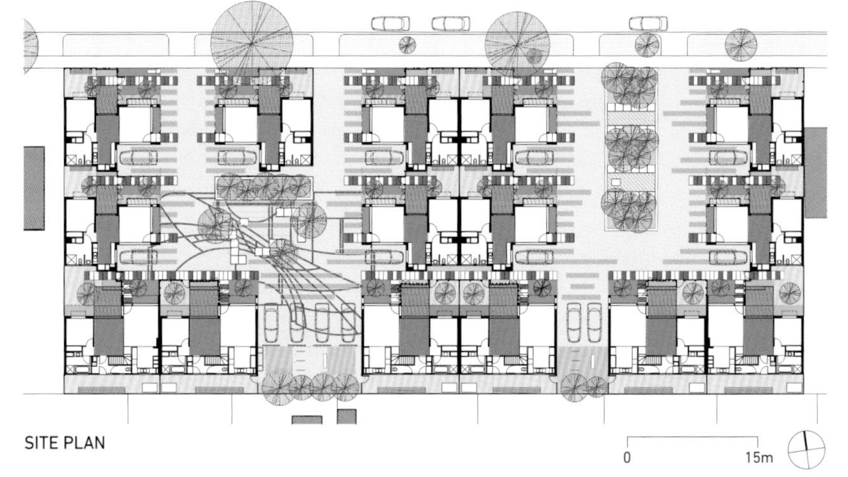

SITE PLAN

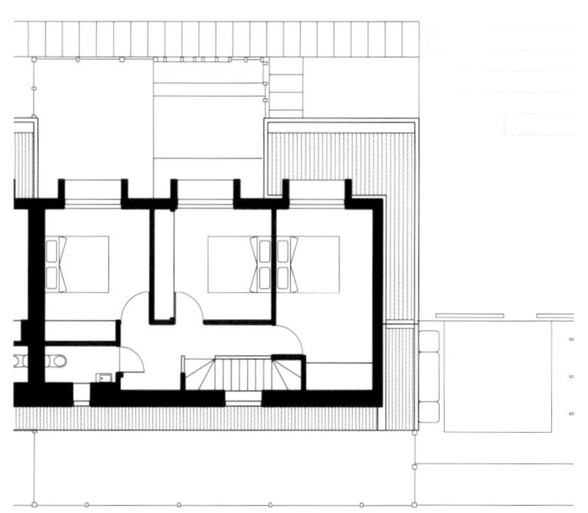

FIRST-FLOOR PLAN (FAMILY UNIT)

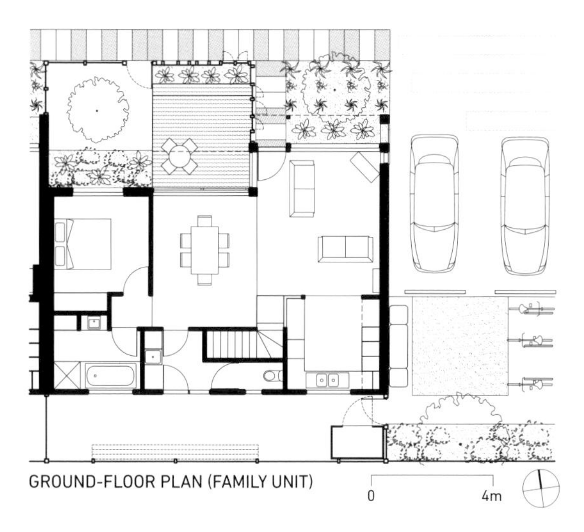

GROUND-FLOOR PLAN (FAMILY UNIT)

格林威治千年村第二期

Greenwich Millennium Village Phase 02

London, United Kingdom

Proctor and Matthews Architects

Phase 02 of the Greenwich Millennium Village includes more than 300 new units designed to meet the needs of the residents in the twenty-first century and respond to changing social patterns. Located on a 32 acre (13-hectare) site on the Greenwich Peninsula near the Millennium Dome in London, it is a mixed waterfront development with townhouses and tall apartment buildings.

Phase 02 is this project's most recent addition to the community and is part of the general Docklands development along the Thames. The initial brownfield site was dormant for the past twenty years, with proposals for sustainable communities from various architects beginning solicitation in 1998. The idea behind the project was that this village was to be built in phases and that each phase would be designed by a different architect. Erskine Tovatt led the first stage of 450 apartment units on a site along the adjacent river. The current second phase, led by Proctor and Matthews Architects, includes 372 dwellings built on three adjacent blocks. A school, a health center, a shopping square, and a parking garage are also included.

The architects emphasized sustainable design and met targets that included a 30 percent reduction in primary energy, a 50 percent reduction in embodied energy, and 50 percent reduction in construction waste. Materials that have been recycled or locally produced were used whenever possible. The use of prefabricated techniques helped speed the process, improve quality, and reduce cost.

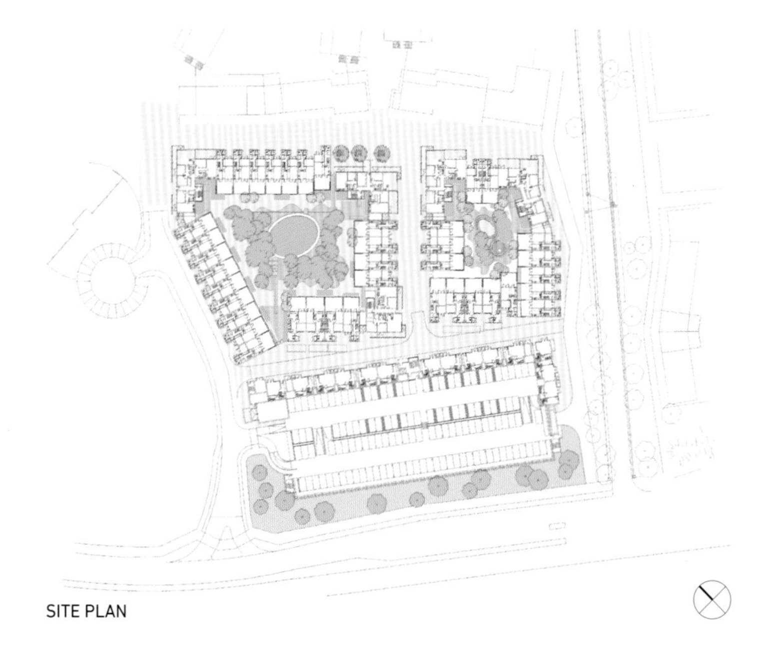

SITE PLAN

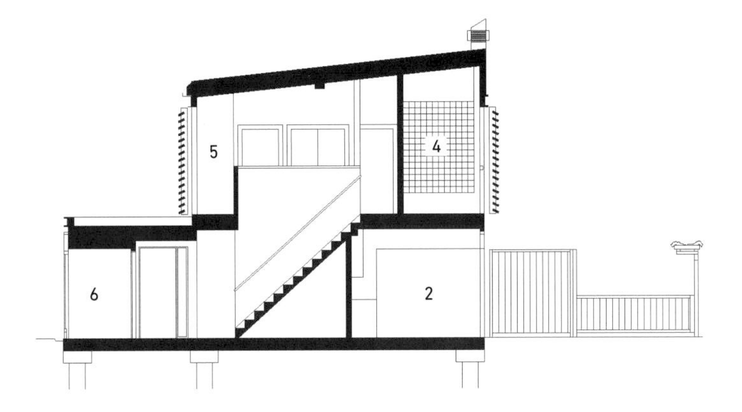

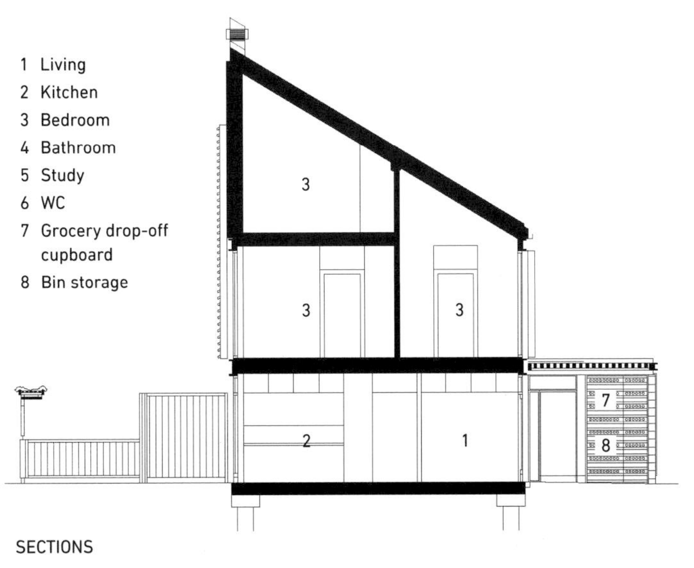

1 Living
2 Kitchen
3 Bedroom
4 Bathroom
5 Study
6 WC
7 Grocery drop-off cupboard
8 Bin storage

SECTIONS

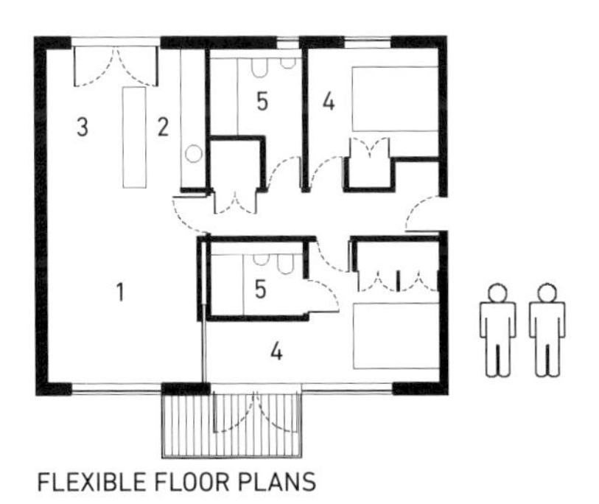

FLEXIBLE FLOOR PLANS

1 Living
2 Kitchen
3 Dining
4 Bedroom
5 Bathroom
6 Study

A community space makes up the heart of the development. The buildings, arranged in clusters, are oriented so that they surround central spaces to support opportunities for social interaction. These clusters, in turn, surround Southern Park. The residences are built on the traditional principle of the London square so that streets and public spaces are scaled to provide a lively, intimate, and secure environment.

The private units are flexible and adaptable—meeting the needs of occupants. The apartment layouts can be changed for a variety of uses. Sliding walls between the living and sleeping areas provide the residents with the option of having an open or cellular plan. The project has since earned distinctions such as the Civic Trust Award and the World Wild life Fund/House Buildings Federation Sustainable New Homes Award.

格林别墅区

Cottages on Greene

East Greenwich, Rhode Island, United States

Union Studio

Cottages on Greene is a privately initiated mixed-housing development in downtown East Greenwich, Rhode Island, and the result of a demand for low-rise and urban-style living. The development showcases that affordable housing for mixed-income residents is attainable through innovative housing methods even during turbulent economic times.

The town of East Greenwich, with its strong historic and cultural values, as well as being close to the cities of Providence and Boston, had become a desirable town for middle- and upper-income residents. By 2010, the housing value had increased and was affected by the high demands for low-density neighborhoods. In addition, the decrease in land supply for future communities also contributed to a limited number of affordable housing initiatives.

Rhode Island's Low and Moderate Income Housing Act of 1991 mandated that at least 10 percent of all housing units should be affordable, which was not the case in East Greenwich. In response to this law, the town approved an affordable housing plan that would allow for the development of a mixed-income community. The 0.85 acre (0.34 hectare) site was then chosen due to its location next to the downtown commercial area.

To help the community appeal to homeowners who wanted a pedestrian urban plan, the design team looked to the historic New England cottage style. The intention was to accentuate high-quality design as opposed to a high volume and so create a link with the old adjacent neighborhoods. The homes in Cottages on Greene borrow from the surrounding neighborhoods in terms of architectural style and specific organization, and were planned in two rows that circulated around communal spaces and existing historic units.

1
B

1 Common green
2 Community gardens
3 Shed
4 Typical private patio
5 Community parking
6 Dumpster enclosure

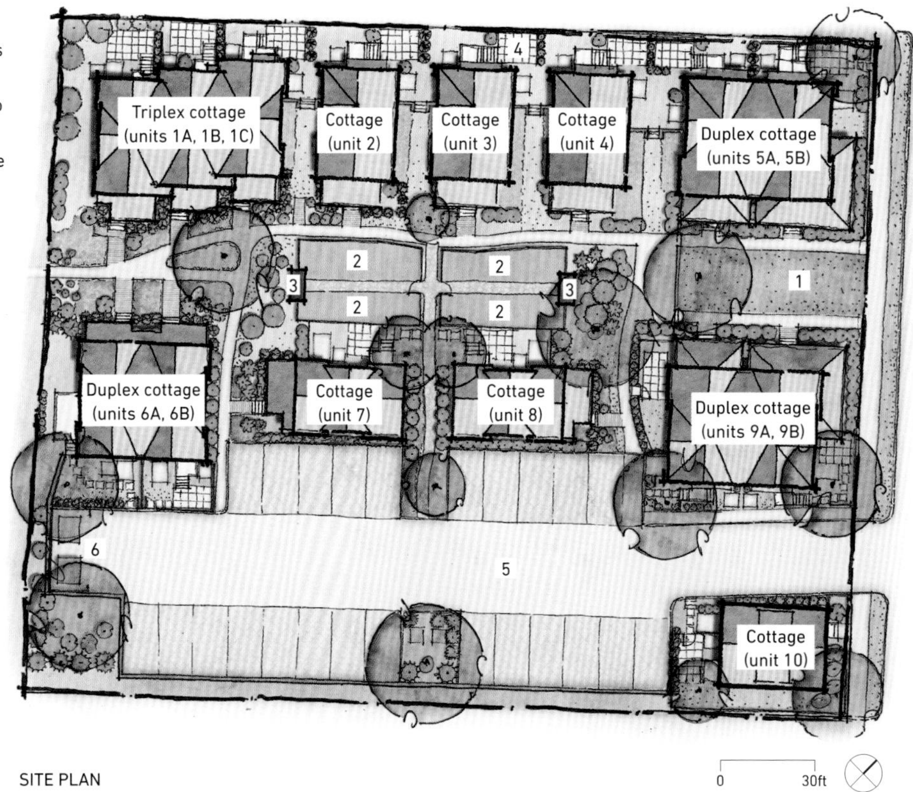

SITE PLAN

With a total of 40 cottage units of various styles including triplexes, duplexes, and single-detached houses, the floor plans are designed to minimize sight lines. As opposed to other developments, low- and moderate-income dwellings are scattered throughout Cottages on Greene, making them fit in with the higher-market-rate units. The design team aimed to provide balance between communal and private spaces and foster a sense of community. In addition, community gardens are also included in the masterplan as well as environmentally conscious techniques such as bioretention and bioswales, permeable pavement for water flow, and an underground stormwater expulsion system. Other sustainable aspects include the pedestrian-friendly location, smaller units, and reduced energy consumption.

第十章：低碳住宅

LOW-CARBON RESIDENCES

With the constant rise in average global temperatures, climate change can no longer be ignored. To reverse the phenomena, designers are introducing buildings that reduce the use of materials with high carbon content and release minimal greenhouse gases (GHG) into the atmosphere. Such buildings are known to have a small ecological footprint. This term can be defined as the metric for calculating the amount of land required to sustain the consumption of natural resources by an individual, community, or country, expressed in equivalent numbers of "planets." It has been calculated that humanity in total uses the equivalent of 1.5 planets for current resource consumption and waste production, revealing that resources are being depleted faster than nature can renew them (GFN 2010). This chapter deals with housing and how it should be designed and built to reduce our carbon footprint.

Current rates of natural resource consumption have created an environment that threatens the earth's biodiversity. Examples include collapsing fisheries, diminishing forest cover, depletion of freshwater systems, and the build up of carbon dioxide emissions (GFN 2010). It is by using tools like the ecological footprint that people will be able to better manage consumption of resources in neighborhoods, to ultimately ensure the survival of the world's ecosystems.

Another useful measure of human impact on the environment is carbon footprint; a subcomponent of the ecological footprint. The carbon footprint is defined as the "total amount of greenhouse gases produced directly and indirectly to support human activities, usually expressed in equivalent tons of carbon dioxide" (Peterson and Rohrer n.d.). Therefore, it is not only carbon dioxide gas emissions that are calculated, but also emissions of other greenhouse gases such as methane and nitrous oxide (Clark 2012).

At present our carbon footprint represents 54 percent of humanity's overall ecological footprint, and this percentage continues to grow (GFN 2012). The burning of fossil fuels is largely responsible for global warming.

Initiatives to minimize the consumption of primary energy and fossil fuels, therefore, will effectively reduce the size of the carbon footprint, and by extension, the ecological footprint as well.

The concept of embodied energy illustrates the idea that current methods for constructing and maintaining homes are significant contributors to the depletion of non-renewable sources of energy. Embodied energy is the amount of energy consumed when building materials are acquired, processed, manufactured, transported, constructed, repaired, and replaced (CSSBI 2013). Embodied energy can be classified into two types: initial and recurring energies. Initial embodied energy is the energy required to manufacture, transportation, and assemble the materials needed in the construction of a building. Recurring embodied energy is the energy needed to repair or maintain a building. Embodied energy, therefore, becomes relevant to the discussion of sustainable design of dwellings in neighborhoods.

The decision to renovate an old house or to build a new one will be influenced by the levels of embodied energy involved. As indicated in chapter six, it is appealing to reuse existing structures because of the cultural heritage conservation values, and also because embodied energy is reduced with building reuse. However, reusing a dwelling will still generate embodied energy, since the house will have to be renovated and adapted to its new use. Furthermore, it has been found that new dwellings are often more energy efficient than older, renovated buildings. For instance, a house that is designed from the beginning with improved insulation will fare better than a dwelling with added insulation. Indeed, it is difficult to add insulation to an existing building as it can consequently create issues with durability (Carpenter 2010).

Many different methods and principles have been developed to minimize embodied energy for new houses. A neighborhood's carbon footprint will be reduced as a result, and these principles may be grouped together as the fundamentals that guide sustainable, low-carbon housing design in contemporary communities.

Houses that address climate change are rapidly becoming more common in the market. First, a house should be efficient in the energy needed for operation. It is important for a home to be airtight and well insulated. Heating and cooling systems are often largely responsible for the energy use in a building, and an airtight house will minimize thermal losses, therefore, wasting fewer energy resources (University of Exeter 2011). Heat can leak through the building envelope or through ventilation ducts. The use of materials with high insulating values will lead to lowering energy consumption. Ventilation systems that incorporate heat recovery also contribute to energy efficiency. In general, the design of a dwelling will focus on keeping heat in and making use of heat gains (RIBA 2013).

Another way to create an energy-efficient house is by manipulating the building form and materials to maximize the use of natural resources to heat and cool the house, instead of relying on mechanical systems. Therefore, a low-carbon house will make use of solar and internal heat gains such as the heat generated by people and equipment, but will still shut out a proper amount of solar gain to avoid overheating. This concept can be taken further by using renewable-energy systems. These houses will generate fewer greenhouse gas emissions (GHG) by using renewable natural resources instead of burning fuels with high carbon-dioxide-emission factors such as coal.

These renewable energies can be harvested through solar, wind, or geothermal district heating systems. Strategic control systems for dwellings are also important, since they will increase the level of comfort in a home and contribute to conserving energy. Default settings for control systems that are set to the lowest energy-consuming level possible and automatically shut down when the system is not needed, will reduce energy consumption. Controls should, however, be intuitive for the homeowner and easy to reach and use to maintain comfort.

Lighting is another important aspect in the design of a low-carbon dwelling. Similar to renewable-energy harvesting strategies, houses can make use of natural lighting to reduce the carbon footprint. Light wells and skylights can let natural light into a house from above, and windows can brighten the interior from the sides. While natural lighting helps to preserve the environment, it also has many health benefits. Strong daylight may contribute to the occupants' wellbeing and mental health. Light that enters a house through skylights will cause less obstruction than windows.

External obstructions and the sun path must be considered when designing windows. For instance, an east-west oriented street will generally preclude sunlight admittance deeper into the street, and it is the top part of the façade facing south that will receive direct light (University of Exeter 2011). In general, direct sunlight on the south façade of a house is the easiest to control, since the sun has the highest altitude when it is due south. As a result, while homes that make use of natural daylight will contribute to sustainability, windows must be strategically placed and oriented to avoid glare.

It is not always possible to provide natural sunlight for a house, and there will be a need to use artificial lighting systems that can still be low carbon. A lighting fixture consumes energy while in use, yet a low energy lighting system can combine natural and artificial lighting strategies. Daylight potential should, therefore, be maximized and artificial lighting systems should use low-energy bulbs. Systems such as occupancy-based control that detect motion will help reduce energy consumption and create greener dwellings.

Materials should also be carefully chosen if attempting to build a low-carbon home. Certain materials have lower embodied energy than others. For example, it takes much less energy to collect and manufacture wood than to produce concrete. However, concrete provides more thermal mass than wood and this can reduce operating energy (Carpenter 2010). Material, therefore, must be cross-compared when designing a low-carbon home for their embodied energy values. Salvaged, recycled, and local materials may also contribute to a reduction in embodied energy in dwelling design.

With global warming, the need to rethink current methods of production and our consumption of natural resources is ever more apparent. Changes in the housing industry offer an opportunity for communities to reduce their carbon footprint, since the construction of buildings and their maintenance is a large contributor to greenhouse gas emissions. Low-carbon residences can form neighborhoods that are leaders in the field of sustainability. Combining initiatives for greener dwellings with effective urban planning policies will help conserve the environment for future generations.

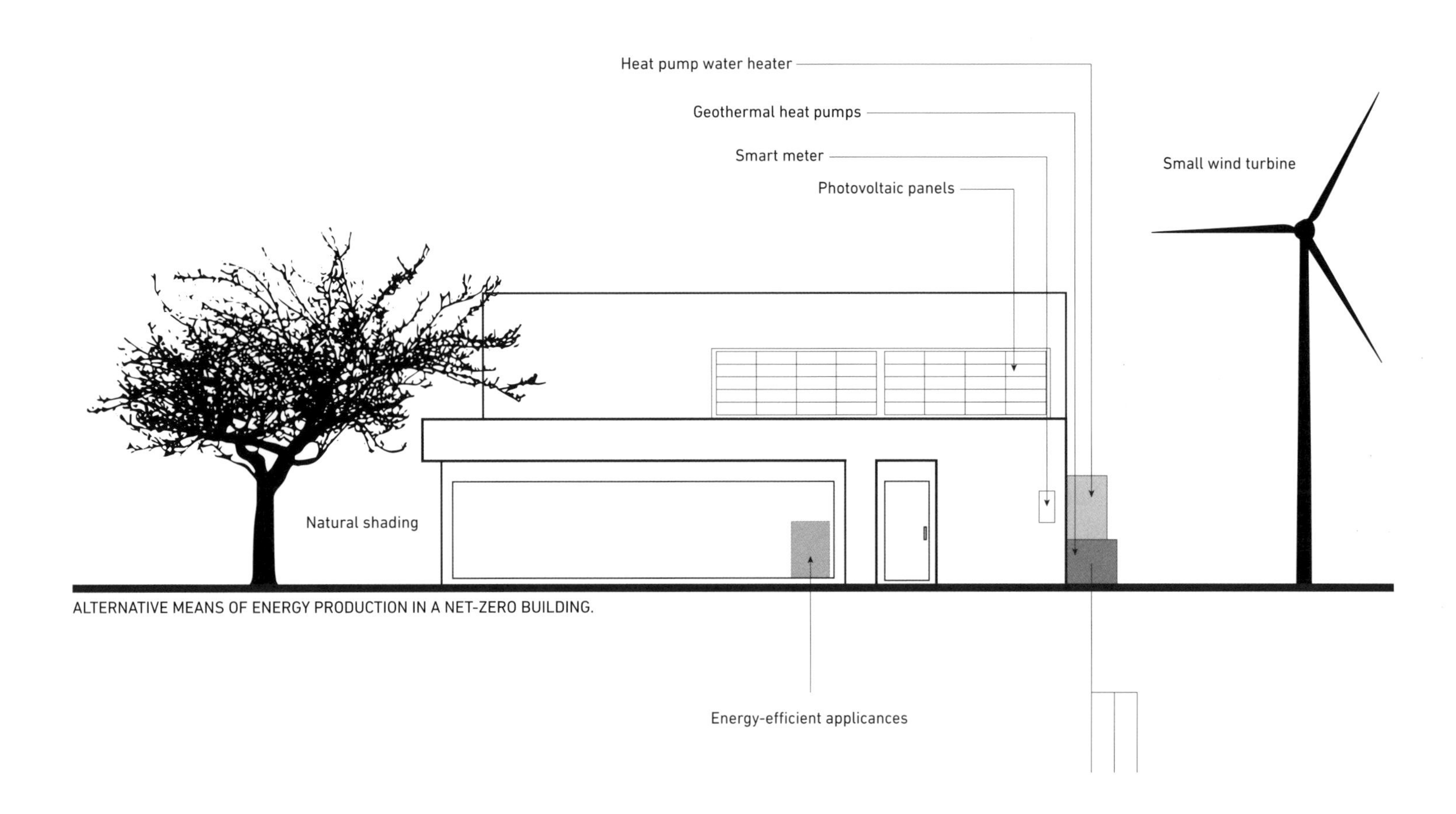

ALTERNATIVE MEANS OF ENERGY PRODUCTION IN A NET-ZERO BUILDING.

哥伦比亚车站微型社区

Columbia Station Micro-community

Seattle, Washington, United States
Dwell Development

The Columbia Station Micro-community is composed of four contemporary residences that will increase in numbers with the development of the next three phases. The first phase was designed by Dwell Development, an award winning design-build firm. Columbia Station had one of the first homes in the Pacific Northwest to meet the Passivhaus standards, an energy efficiency label more difficult to achieve than LEED.

The development is located in a charming historic district on an infill lot within the Seattle housing authority's mixed-use, mixed-income, Rainer Vista development. The four small housing units are either two-bed/two-bath or three-bed/three-and-a-half bath models. The neighborhood is diverse and vibrant with a farmers' market, thriving arts scene, ample green space, and public services such as the light rail station and public transportation within walking distance. Residents of Columbia Station are also able to enjoy the benefits of a walkable neighborhood.

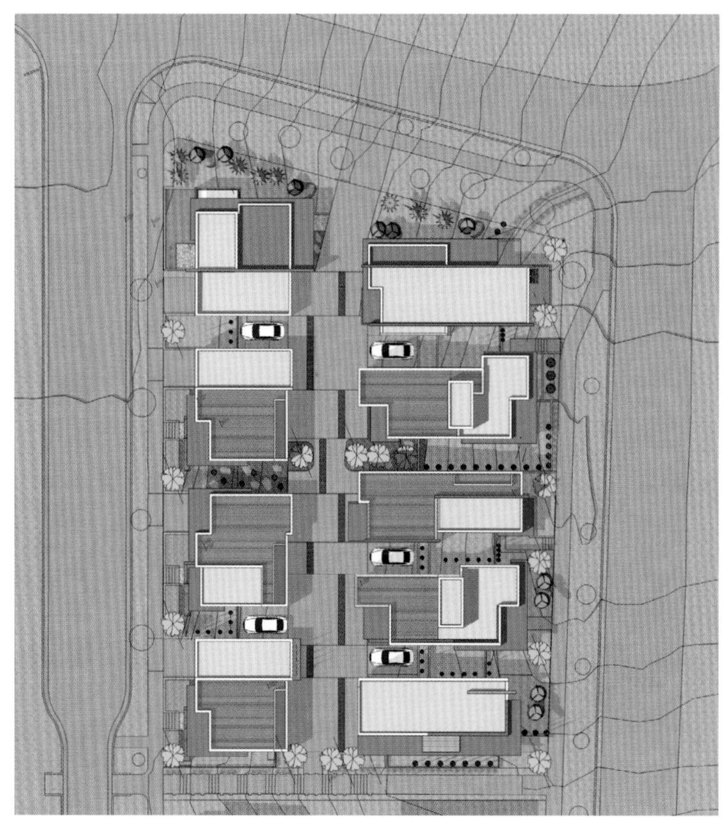

SITE PLAN

S Oregon St
NEW HOME

4459
4453

Phase one of the Columbia Station Micro-community has received a green five-star verification. To obtain the Passivhaus standard, the project must be able to reach thermal comfort levels by post-heating or post-cooling fresh air without necessitating recirculation. Sustainable features employed include technology such as the use of super-tight fiberglass insulation, recycled-glass counter tops, dual-flush toilets, low-flow water fixtures, energy star appliances, triple-plane windows, heat-recovery ventilators, re-milled pine flooring from reclaimed telephone poles, low VOC paints, finishes, adhesives, and also sustainable landscaping, such as rooftop gardens and rainwater collection systems.

The units in phase one of the project were purchased before the project's completion. Now there is much anticipation and demand for phase two, which will include the first built-on-spec Passivhaus in the United States.

4440

奥斯丁SOL社区 (1.0版)

SOL Austin (V1.0)

Austin, Texas, United States

KRDB

The SOL Austin (V1.0) project is located 3 miles (4.5 kilometers) east of downtown Austin, Texas, a short walk from Govalle Park and Lady Bird Lake. The design is part of a comprehensive approach to sustainable development that consists of a green community with net-zero homes that are well linked to the larger urban context. The "version 1.0" of this master plan is made up of 40 units.

The macro-strategy of the project was to minimize the environmental impact and not change the existing urban infrastructure of the site. To create a more advantageous balance between public and private spaces, the lots are reduced in size. The homes do not have a uniform streetscape, but varied setbacks for a more interesting and heterogeneous public realm.

The driveways are shared, which helps to reduce the amount of surface cover and vehicle presence. There are also many biking, running, and walking pathways. There are ample community green spaces and all landscaping employed is chosen to be drought resistant, low-impact vegetation so little or no water is needed for irrigation aside from typical rainfall.

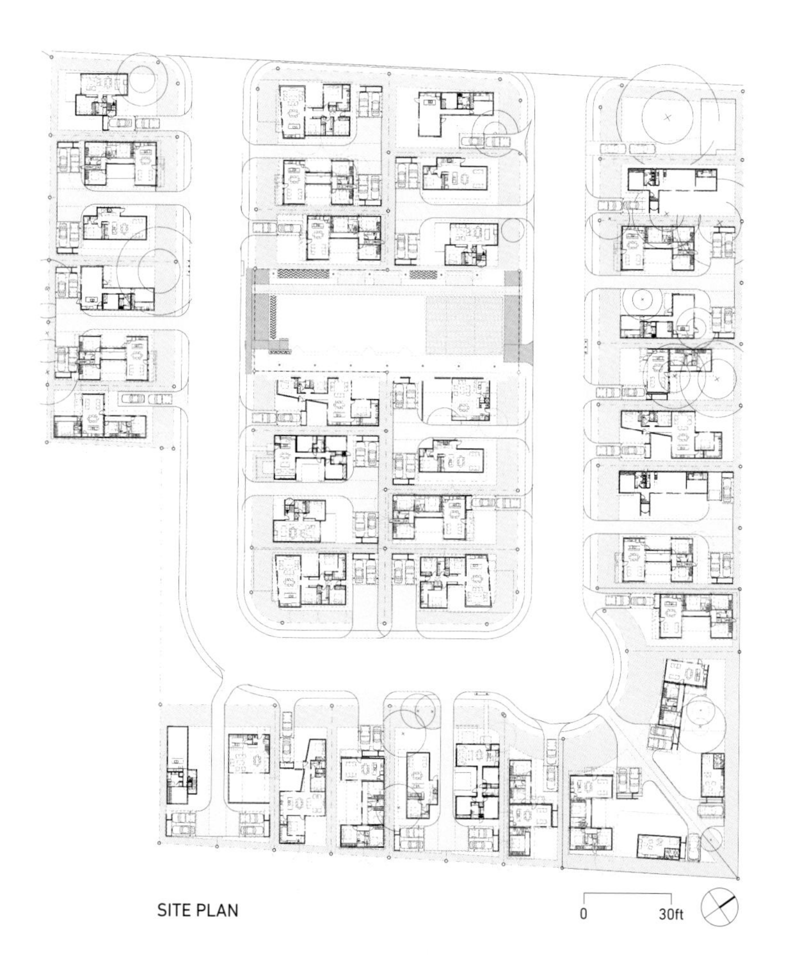

SITE PLAN

To achieve its net-zero home status, the designers of SOL Austin residences employ many sustainable techniques including the use of high-quality windows, 2x6 wood frame construction, geothermal Ventilation Heating and Air Condition (HVAC), and solar energy systems. The majority of the windows face south and north rather than east and west in combination with operable windows for passive ventilation and day lighting. This excellent passive design strategy minimizes energy usage in both summer and winter. Providing thermally efficient high- quality windows is critical in the sustainability of the home, where every "break" in the exterior signifies an increased stress on the HVAC system. Hence, Gerkin Rhino windows were used, which are a low-e, double-pane, with a thermally broken aluminum frame featuring Cardianl 366 glass that exceeds the energy star guidelines for Austin's climate.

The use of 2x6 studs for all exterior walls and spray-in Icynene insulation when constructing the internal framework were found to have as high an R-value and envelope tightness as that of structurally insulated panels (SIP). Using the earth as a heatsink, geothermal HVAC heating and cooling systems that require only half of the conventional energy were used, plus roof-top photovoltaic systems customized in size according to the occupancy level. SOL is the first development in Austin to use the method of sub-grade bio-filtration that directs runoff from the streets, sidewalks, and driveways into the sedimentation pond and subsequent filtration system. Each home also has the necessary equipment for rainwater collection. With all these energy saving strategies, the total energy demand for each house is reduced by 50 percent.

As part of the project's holistic sustainability and affordability model, 40 percent of the units were affordable housing. Eight units were for sale to first time homebuyers making 80 percent of the median family income or less, and eight units are rental housing for people making 30 to 60 percent of the median family income.

甜水光谱社区

Sweetwater Spectrum Community

Sonoma, California, United States

Leddy Maytum Stacy Architects

The Sweetwater Spectrum Community designed by Leddy Maytum Stacy Architects is devoted to support housing for adults with autism, offering them a comfortable and safe environment. This 2.8 acres (1.13 hectares) site with four four-bedroom units provides a permanent home for sixteen adults and their supporting staff. Each resident is equipped with their own bedroom and bathroom, common areas in each home, a community center with exercise and activity spaces, teaching kitchen, a large therapy pool and spas, an urban farm, orchard, and greenhouse. The Sweetwater Spectrum Farm allows residents and program volunteers to grow and sell their own produce such as fruit, vegetables, and herbs.

Although autism is one of the fastest growing developmental disabilities in the United States, there were minimal residential options. In 2009, a group of families who had children with austism, autism-care professionals, and community leaders founded Sweetwater Spectrum—a non-profit organization to develop appropriate, high-quality, and long-term housing for adults who have been diagnosed on the autism spectrum. This new community addresses the various needs of the individuals with autism spectrum disorders to maximize development, independence, and overall wellbeing.

The development was built in proximity to the historic Sonoma Town Square, public transportation, cycling trails, shopping, dining, and entertainment venues. The project is designed to LEED Gold level certification requirements and residents enjoy reduced energy consumption, long-term operating cost efficiency, and a healthier indoor atmosphere. As a participant in the net-zero energy pilot program, Sweetwater Spectrum combines electrical and mechanical aspects that maximize energy efficiency such as a water-conservation program, a renewable and recyclable program, and an energy-efficient community design. The water conservation program includes the use of drought-tolerant plants, on-site stormwater management, water-efficient plumbing fixtures, and on-site well water for cultivating the urban farm and orchards. The community was built using renewable and non-toxic building materials, including Forest Stewardship Council certified lumber, low volatile organic compound materials, the use of non-formaldehyde adhesives, and non-toxic building materials.

Aspects of this energy-efficient design include the use of solar water heating, low E-value insulated windows, high R-value insulation, natural daylight and sun control, and the use of solar heating and cooling slabs with high-efficiency and low-velocity air ventilation. Other sustainable energy tactics include using an urban infill site, minimizing changes to the natural surroundings during building and daily community operations, and an indoor air quality management system. Furthermore, solar reflective paving and "cool" roofs are also implemented in the design in order to reduce heat island effect. In preparing for the use of photovoltaic solar electric (PV) roof panels, the design team focused on optimal roof pitch and orientation, ventilation layout, and specific roof load bearings.

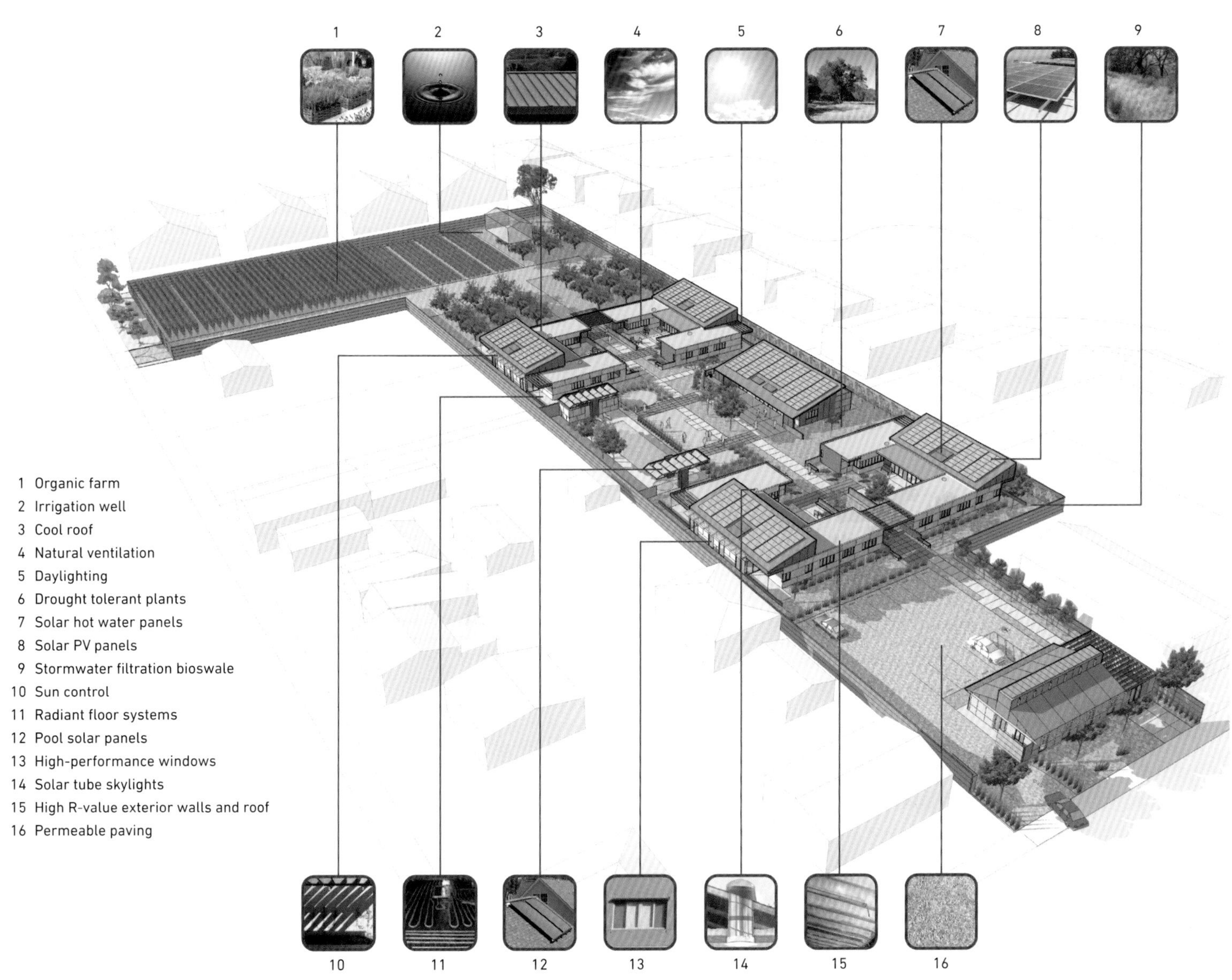

SUSTAINABILITY IMPLEMENTATION DIAGRAM

同胞绿色社区

Paisano Green Community

El Paso, Texas, United States

WORKSHOP8

Paisano Green Community, designed by WORKSHOP8, is an infill development surrounded by industrial and civic uses. Buildings placed on the edge of the site emphasize boundaries while creating a secluded and central communal garden inside. Covering an area of 4.2 acres (1.40 hectares) and providing a total of 73 housing units of varying styles, this community is devoted exclusively to seniors. Paisano is defined as the first senior community that is affordable, net-zero, fossil-fuel free, and LEED Platinum.

The 73 units in six buildings include eight two-bedroom units, three courtyard units, nine single-room-occupancy townhomes, and 53 one-bedroom units. WORKSHOP8's design attempted to create a project that would promote a healthier lifestyle, while being environmentally conscious and fostering spontaneous interpersonal relationships. To address energy issues, the design paid particular attention to site-planning and building-design strategies related to solar orientation. The architectural expression of the buildings includes extended overhangs in parts facing south, limits north- and east-facing windows, and doesn't feature windows facing west. All housing units allow residents to have south- and north-facing fiberglass windows that are fully operable and can be opened so that fresh cool breezes can be experienced indoors. The buildings themselves have a very tight envelope and insulation, providing minimal natural air changes per hour.

Due to El Paso's harsh climate, Paisano Green was faced with the challenge of providing significant levels of heating and cooling while being energy efficient. To tackle this issue, the project utilizes a ductless, mini-split, heat pump system. Water heating is achieved by air-source, heat-pump water heaters that take energy from the air within the unit, inject this energy into the water, and eject cooler air back into the interior. This system is known to be three times more efficient than the standard electrical-resistance water heater system. All housing units include an Energy Recovery Ventilator (ERV) heat exchange units. These are designed to economically capture energy in the air and transfer it to incoming fresh air.

DO NOT
ENTER
A
B
C
D

Other strategies to conserve energy include minimizing manipulation of the surrounding environment during building and day-to-day operations, use of a solar chimney, use of a stormwater management system and drought tolerant plants.

These various energy efficient systems, integrated with passive energy design strategies, equate to minimal use of the installed PV panels leading to net-zero status. All aspects of these sustainable design strategies provide optimal comfort for senior residents, and minimize operating and maintenance costs.

1 Entry plaza
2 Community building
3 Canopy roof over community building terrace
4 Plaza
5 Boone Avenue residences
6 South shade structure
7 Tapestry garden
8 Picnic area
9 Courtyard 1
10 Courtyard 2
11 Courtyard 3
12 Flat building A
13 Flat building B
14 Flat building C
15 Flat building D
16 Visitor parking
17 Resident parking
18 Wind turbines
19 Canopy wall
20 Entry pavilion
21 PV panels

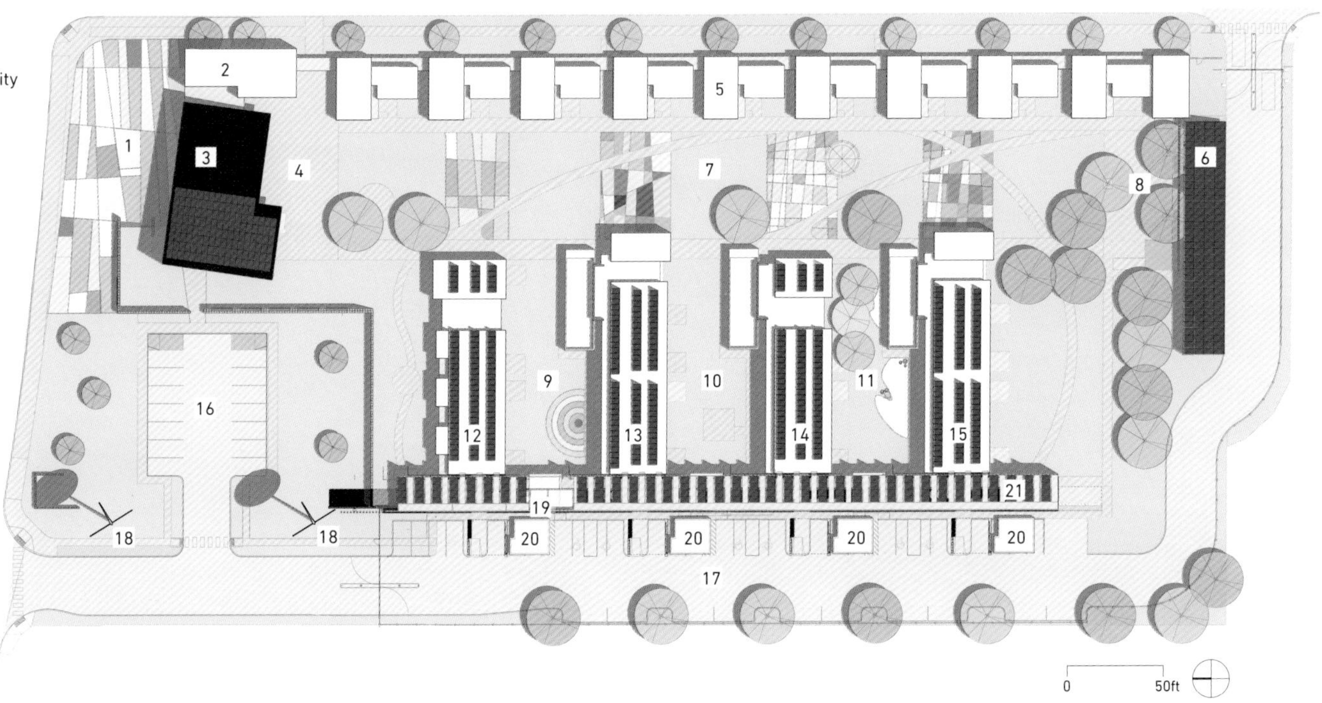

SITE PLAN

C

z之家社区

zHome

Issaquah Highlands, Washington, United States
David Vandervort Architects

zHome community is a transit-oriented development with ten net-zero energy townhomes on a 0.4-acre (0.16-hectare) site near the entrance to the Issaquah Highlands neighborhood. The place is one of many that are shifting the way architects and engineers plan and develop such places.

The architects were concerned with having easily accessible amenities while providing residents with privacy. Function was always planned and tied with the mission of creating a highly sustainable neighborhood using tactics such as providing residents with clean and renewable energy from heat pumps, using photovoltaic (PV) panels, while implementing passive-energy design techniques.

One key feature in zHome is an interior solar courtyard which is defined as a place that connects both the social and ecological issues of the project. The reason behind introducing this space was a desire to create a community where residents could have a comfortable social area. This draws attention to the interior courtyard, street facing views, and engages public interactions. Creating private spaces for inhabitants was also a key feature.

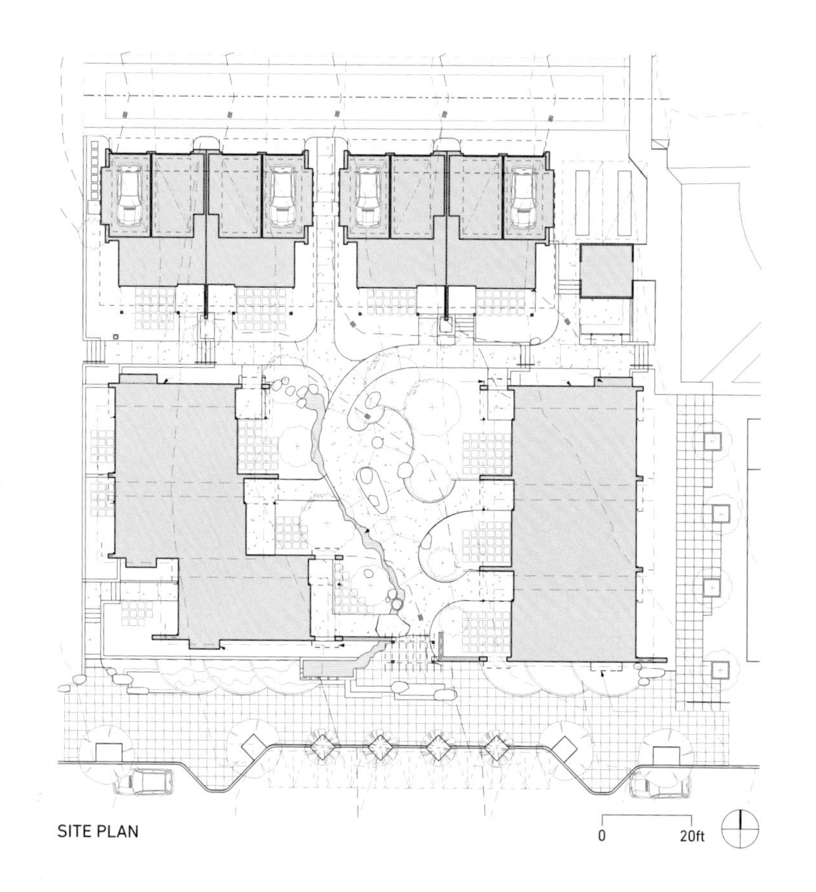

SITE PLAN

zHome

For example, each housing unit is equipped with a spacious living area connected to private decks and porches all on the second floor. These parts of the dwellings serve as an opening that can either combine with the public spaces or be enclosed entirely, providing optimal flexibility.

Optimizing clean, fresh air, zHome is designed so that automobiles operate outside of the development with specialized garages and vehicle amenities all located on the northern perimeter. This tactic gives residents security, clean air, and limits traffic congestion.

It enables residents to engage in spontaneous encounters and use other forms of transportation within the community such as walking and cycling. Harvesting, storing, and sharing rainwater among the units is a key feature in the design of zHome. This is achieved by collecting stormwater and filtering it, which minimizes the overflow onto paths and public spaces. Activities such as recycling, waste disposal, and gardening at the communal garden can all be found in one location, optimizing the sharing of utilities and enriching the quality of the community both socially and physically.

Additional, specific design techniques include: optimizing roof areas for solar photovoltaic panels, using high-quality windows and glazing, and allowing for optimal ventilation and thermal mass for increased insulation. One important characteristic to note is that the designers did not let these techniques overrule the architectural language of the project. ZHome was recognized by the International Living Future Institute with its Net Zero Energy Building Certification and Petal Certification under its Living Building Challenge program in 2013.

致谢

Acknowledgments

The design and planning of homes and communities was the focus of my work for many years. It included collaboration with colleagues and assistants who contributed to the ideas expressed here. I have attempted to remember and acknowledge them all. My apologies if I have mistakenly omitted the name of someone who contributed to the ideas, text or illustrations that have been included in this book. I will do my best to correct any omission in future editions.

The book could not have been written without the contributions by a team of highly dedicated students. Lisa Chow and Roxanne Turmel contributed to the background research and writing of the chapters.

Sara D'Amato played a key role in finding the projects and describing them. Her dedication, hard work and interest in the subject are much appreciated. She also helped to compile the bibliographies with utmost attention and created some of the illustrations that accompany the chapters.

David Auerbach, Charles Gregoire, Caroline Pfister, and Rose Deng were also instrumental in finding the projects and describing them. Special thanks to all the design firms and the photographers whose work is included here for making the material available.

I would like to express my gratitude to Paul Latham, Honorine LeFleur, Rod Gilbert, Hannah Jenkins, and Bethany Patch at Images Publishing for ushering the book in and for their patience and guidance in seeing it through.

Thanks to McGill University's School of Architecture where the genesis of the ideas expressed here and my own research were carried out.

Finally, my heartfelt thanks and appreciation to my wife Sorel Friedman, Ph.D., and children Paloma and Ben for their love and support.

章节参考书目

Bibliography for Chapters

Atkinson, Rick. "Houses for Multigenerational Families." Buildipedia, 2013. Last accessed April 30, 2015 from http://buildipedia.com/at-home/design-remodeling/houses-for-multigenerational-families

Bartlett, Liz. "Introducing Future Communities." *Home Page, Young Foundation*. 2009. Last accessed April 30, 2015 from http://www.futurecommunities.net/

Bhatt, Vikram. "Making the Edible Landscape: A Study of Urban Agriculture in Montreal." Minimum Cost Housing Group. McGill. 2005. Last accessed April 30, 2015 from http://www.mcgill.ca/mchg/pastproject/el

Bollerud, Erica. "Heat Island Effect." *EPA, Environmental Protection Agency*. March 7, 2013. Last accessed April 30, 2015 from http://www.epa.gov/hiri/

British Columbia. *Ministry of Forests, Lands, and Natural Resource Operations*. "Introduction to Heritage and Sustainability." November 29, 2012. Last accessed April 30, 2015 from http://www2.gov.bc.ca/gov/topic.page?id=D6F6003B73D14A26BB68E3C824FDDB83

Burke, Denver. "Is District Heating the Way Forward for Renewable Energy?" *RenewableEnergyWorld.com*. December 12, 2012. Last accessed April 30, 2015 from http://www.renewableenergyworld.com/rea/blog/post/2012/12/district-heating-can-be-seen-as-the-way-forward-for-renewable-energy

Canadian Sheet Steel Building Institute (CSSBI). "Embodied Energy in Buildings." 2013. Last accessed April 30, 2015 from http://cn-sbs.cssbi.ca/embodied-energy-in-buildings

Carpenter, Stephen. "How important is embodied energy?" June 1, 2010. Last accessed April 30, 2015 from http://www.building.ca/news/how-important-is-embodied-energy/1000378217/

Clark, Duncan. "What's a carbon footprint and how is it worked out?" *The Guardian*. April 4, 2012. Last accessed April 30, 2015 from http://www.theguardian.com/environment/2012/apr/04/carbon-footprint-calculated

Cheng, Aaron. "High Density vs. Low Density." *High Density vs. Low Density.* 2007. Last accessed April 30, 2015 from http://www.chengfolio.com

CIT Energy Management AB. "Success Factors in Solar District Heating." *Intelligent Energy Europe.* December 2010. Last accessed April 30, 2015 from http://www.solar-district-heating.eu/Portals/0/SDH-WP2-D2-1-SuccessFactors-Jan2011.pdf

City of Pointe-Claire. "Planning Program Chapter 3: Proposed Vision, Major Land Use Designations, Transportation and Detailed Planning Sectors." November 15, 2010. Last accessed April 28, 2013 from http://www.ville.pointe-claire.qc.ca/en/planning/

"Construction." *Sector, Biodiversity and Wildlife Conservation. The Business & Biodiversity Resource Centre*. 2007. Last accessed April 30, 2015 from http://www.businessandbiodiversity.org/construction.html

Council of Europe. *Manifesto for a new urbanity: European Urban Charter II*. Council of Europe Publishing. March 2009.

Current, Dean. "Selection Trees and Shrubs in Windbreaks." *Extension, University of Minnesota.* Fall 2011. Last accessed April 30 2015 from http://www.extension.umn.edu/environment/agroforestry/components/selecting-trees-and-shrubs-in-windbreaks.pdf

Dalkia Canada. "District Heating and Air-Conditioning Systems." 2009. Last accessed April 30, 2015 from http://www.dalkia.bg/en/solutions/businesses/heating-and-cooling-systems/

"Determining the MP 2003 Value." *2008 Development Baseline Definition – A Guide for the Property Industry*. Dec. 2007. Last accessed April 30, 2015.

District of Saanich. "Adaptable Housing Frequently Asked Questions." December 12, 2012. Last accessed April 30, 2015 from http://www.saanich.ca/business/adaptable/adaptablefaq.html

EA Energy Analyses. "Wind and heat: a pilot project on district heating from wind turbines and heat pumps." September 15, 2011. Last accessed April 30, 2015 from http://www.ea-energianalyse.dk/subpages/929_Wind_and_heat_a_pilot_project_on_district_heating.html

Egan, John. "Skills for Sustainable Communities." *The Egan Review. Eland House*. April 2004. Last accessed April 30, 2015 from http://dera.ioe.ac.uk/11854/1/Egan_Review.pdf

Enger, Susan. *Planning for Parks, Recreation, and Open Space in Your Community*. Washington: Growth Management Services, February 2005. Print.

Enwave Energy Corporation. "Deep lake water cooling." 2011a. Last accessed April 30, 2015 from http://www.enwave.com/district_cooling_system.html

Enwave Energy Corporation. "District heating system." 2011b. Last accessed April 30, 2015 from http://www.enwave.com/district_heating_system.html

Euroheat & Power. "District heating and cooling." 2013. Last accessed April 30, 2015 from http://www.euroheat.org/District-heating-cooling-4.aspx

Euroheat & Power. "District Heating in Buildings." 2011. Last accessed April 30, 2015 from http://www.euroheat.org/Files/Filer/documents/Publications/District%20Heating%20in%20buildings_final.pdf

Friedman, Avi. *Homes Within Reach: A Guide to the Planning, Design and Construction of Affordable Homes and Communities*. John Wiley & Sons Inc., New Jersey, 2005.

Friedman, Avi. "Farming in Suburbia." Open House International, Vol. 32, no. 1. March 2007.

Friedman, Avi. *Fundamentals of Sustainable Dwellings*. Island Press, 2012.

Friedman, Avi. *Town and Terraced Housing; For Affordability and Sustainability*. Routledge, London, UK, 2011.

Gilmer, Maureen. "Landscape Windbreaks." *Landscape Network*. 2010. Last accessed April 30, 2015 from http://www.landscapingnetwork.com/landscaping-ideas/windbreaks.html

Global Footprint Network (GFN). "Carbon Footprint." Last modified July 17, 2012. Last accessed April 30, 2015 from http://www.footprintnetwork.org/en/index.php/GFN/page/carbon_footprint

Global Footprint Network (GFN). "Footprint Basics – Introduction." November 3, 2010. Last accessed April 30, 2015 from http://www.footprintnetwork.org/en/index.php/GFN/page/basics_introduction/

"Green Parking Lot Resource Guide." *Stream Tea Mok. Environmental Protection Agency*, February 2008. Last accessed April 30, 2015 from http://www.streamteamok.net/Doc_link/Green%20Parking%20Lot%20Guide%20%28final%29.PDF

Green, Penelope. "Under One Roof, Building for Extended Families." *The New York Times*. November 29, 2012. Last accessed April 30, 2015 from http://www.nytimes.com/2012/11/30/us/building-homes-for-modern-multigenerational-families.html?_r=0

Gromicko, Nick. "Building Orientation for Optimum Energy." *Inspecting the World, International Association of Certified Home Inspectors, Inc*. 2013. Last accessed April 30, 2015 from http://www.nachi.org/building-orientation-optimum-energy.htm

Harvey, L. D. Danny. *A Handbook on Low-Energy Buildings and District-Energy Systems: Fundamentals, Techniques, and Examples*. Earthscan. London: Sterling, 2006.

"Highland's Village Garden." *Highland's Garden Village*. 2013. Last accessed April 30, 2015 from http://www.highlandsgardenvillage.net/

Hood, Lyndon. "Action against Tobacco, Pollution, and Lack of Exercise." *Scoop Independent News*. Last accessed April 30, 2015 from http://www.scoop.co.nz/stories/WO1106/S00460/action-against-tobacco-pollution-and-lack-of-exercise.htm

Kackar, Adhir, and Ilana Preuss. *Creating Great Neighborhoods: Density in your Community*. Washington: The Voice for Real Estate, 2003.

Lea, Keya. "Orientation / South Facing Windows." *Green Passive Solar Magazine*. March 13, 2012. Last accessed April 30, 2015 from http://greenpassivesolar.com/passive-solar/building characteristics/orientation-south-facing-windows/

Linderoth, Terry. "Multi-Family Housing." *Linderoth Associates, Inc*. 2010. Last accessed April 30, 2015 from http://www.linderoth.com/home-multi-family-1.html

Mannvit Engineering. "What is Geothermal?" 2013. Last accessed April 30, 2015 from http://www.mannvit.com/GeothermalEnergy/WhatisGeothermal

Meinhold, Bridgette. "Net-Zero Yin Yang House Soaks up the Sun's Rays in Venice Beach, California." *Inhabitat*. March 12, 2012. Last accessed April 30, 2015. http://inhabitat.com/net-zero-yin-yang-house-soaks-up-the-suns-rays-in-venice-beach-california/

"Model Right to Farm By-Law." *Smart Growth/Smart Energy*. Last accessed April 30, 2015 from http://www.mass.gov/envir/smart_growth_toolkit/bylaws/Right-to-Farm-Bylaw.pdf

Nash, Andrew. "Improving Public Transport Efficiency." *Vienna Transport Strategies*. March 27, 2010. Last accessed April 30, 2015 from http://andynash.com/projects/

Natural Resources Defense Council (NRDC). "Food miles: How far your food travels has serious consequences for your health and the climate." November 2007. Last accessed April 30, 2015 from http://food-hub.org/files/resources/Food%20Miles.pdf

Nordic Folkecenter for Renewable Energy. "District heating can now store wind energy." September 2010. Last accessed April 30, 2015 from http://www.folkecenter.net/gb/rd/power-balancing/district-heating/

"Our Areas of Expertise." *Design Council* – CABE. Last accessed April 30, 2015 from http://www.designcouncil.org.uk/our-work/CABE/Our-big-projects/

Palmer, Jasmine. "The Adaptable House." *Commonwealth of Australia*. 2010. Last accessed April 30, 2015 from http://www.yourhome.gov.au/

Paster, Pablo. "Ask Pablo: What's The Impact Of Imported Tropical Fruit?" TreeHugger, MNN Holdings, Inc. June 23, 2011. Last accessed April 30, 2015 from http://www.treehugger.com/green-food/ask-pablo-whats-the-impact-of-imported-tropical-fruit.html

Peterson, Ann-Kristin, and Rohrer, Jürg. "What is a carbon footprint – definition." n.d. Last accessed April 30, 2015 from http://timeforchange.org/what-is-a-carbon-footprint-definition

Piedmont-Palladino and Mennel. *Green Community*. Chicago: American Planning Association, 2009.

Rabinowitz, Phil. "Improving Parks and Other Community Facilities." *The Community Tool Box. University of Kansas*. 2013. Last accessed April 30, 2015 from http://ctb.ku.edu/en/table-of-contents/implement/physical-social-environment/parks-community-facilities/main

R., Angel. "Taking The Bus (vs) Compared To Driving A Car – Its Effects on Pollution, Global Warming and our Health." July 23, 2013. Last accessed April 30, 2015 from http:// hubpages.com/hub/Gas-vs-Bus

"Road Traffic and Air Pollution." *Health Canada*. April 27, 2011. Last accessed April 30, 2015 from http://healthycanadians.gc.ca/healthy-living-vie-saine/environment-environnement/outdoor-air-exterieur/traf-eng.php

Rodriguez, Luis. "Making Our Communities More Liveable: Examples from Germany and Scandinavia." *Sustainable Cities Collective. Siemens*. Last accessed April 30, 2015 from http://sustainablecitiescollective.com/luis-rodriguez/131416/road-map-making-our-communities-more-liveable-examples-germany-and-scandinavia

Roseland, Mark. "Sustainable Community Development: Integrating Environmental, Economic, and Social Objectives." *Progress in Planning*. Pergamon, 2000. Last accessed April 30 2015 from http://socialeconomyhub.ca/sites/socialeconomyhub.ca/files/Sustainable%20Community%20Development%20-%20integrating%20environmental,%20economic%20and%20social%20objectives_0.pdf

Royal Institute of British Architects (RIBA). "Principles of low carbon design." 2013. Last accessed April 15, 2015 from http://www.architecture.com/files/ribaholdings/policyandinternationalrelations/policy/environment/2principles_lc_design_refurb.pdf

RUAF Foundation. "What is Urban Agriculture?" Resource Centres on Urban Agriculture & Food Security. 2013. Last accessed April 30, 2015 from http://www.ruaf.org/urban-agriculture-what-and-why

Shackell, Aileen, and Robin Walter. *Greenspace Design for Health and Wellbeing*. Edinburgh: Forestry Commission, 2012.

Smallidge, Peter. "Frequently Asked Questions." *Cornell Sugar Maple Research and Extension Program. New York State Agricultural Experiment Station*. Last accessed April 30, 2015 from http://maple.dnr.cornell.edu/FAQ.htm

Smit, Jac, et al. *Urban Agriculture: Food, Jobs, and Sustainable Cities*. The Urban Agriculture, 2001.

Snyder, Marcie. "The Role of Heritage Conservation Districts in Achieving Community Improvement." University of Waterloo. 2008. Last accessed April 30, 2015 from http://uwspace.uwaterloo.ca/bitstream/10012/3801/1/MSnyder_final%20copy.pdf

Space, CABE. "Community Green: Using Local Spaces to Tackle Inequality and Improve Health." *Research Summary. Commission for Architecture and the Built Environment*. 2010. Last accessed April 30, 2015 from http://www.openspace.eca.ed.ac.uk/pdf/appendixf/OPENspacewebsite_APPENDIX_F_resource_2.pdf

Statistics Canada. "Census families by number of children at home, by province and territory (2011 Census)." *Government of Canada*. February 13, 2013. Last accessed April 30, 2015 from http://www.statcan.gc.ca/tables-tableaux/sum-som/l01/cst01/famil50a-eng.htm

Strickler, Sherri. "How working from home is contributing to our sustainability efforts: EarthSmart @ Work." June 12, 2012. Last accessed April 30, 2015 from http://blog.van.fedex.com/blog/earthsmart-at-work

"Sustainable Communities." *EPA, Environmental Protection Agency*. December 17, 2011. Last accessed April 30, 2015 from http://www.epa.gov/sustainability/

"Sustainable Community Development Guidelines." *Park Downsview Park Inc.* December 2007. Last accessed April 30, 2015 from http://www.downsviewpark.ca/

Tesei, Peter. "Residential Green Space-Proposed Zoning Regs." *Town of Greenwich, Connecticut*. 2013. Last accessed April 30, 2015 from http://www.greenwichct.org/News/Planning_and_Zoning/RESIDENTIAL_GREEN_SPA_SPACE_PROPOSED_ZONING_REGS/

Tumlin, Jeffrey. *Sustainable Transportation Planning: Tools for Creating Vibrant, Healthy, and Resilient Communities*. John Wiley & Sons Inc., New Jersey, 2012.

Turner, Michael. "The Historic Urban Landscape Approach and the Future of Urban Heritage." *Wiley*. 2013. Last accessed April 30, 2015 from http://www.wiley.com/WileyCDA/WileyTitle/productCd-1118383982,subjectCd-AR60.html

University of Exeter. "Low Carbon Building Design: Course Notes." March 2011. Last accessed April 30, 2015 from http://www.exeter.ac.uk/news/events/details/index.php?event=451

Urban Fish Farmer (UFF). "Urban Fish Farmer". 2013. Last accessed April 30, 2015 from http://urbanfishfarmer.com

项目参考书目

Bibliography for Projects

1.1 Civano

Buntin, Simmons. "Unsprawl case study: Community of Civano, Arizona." *Terrain.org*. Last accessed August 5, 2015 from http://www.terrain.org/unsprawl/5/

1.2 Prairie Crossing

Prairie Crossing. "About Prairie Crossing." *Prairie Crossing – A Conservation Community*. 2009. Last accessed July 27, 2015 from http://www.prairiecrossing.com/pc/site/about-us.html

Tigerman-McCurry Architects. "Prairie Crossing." *Tigerman-McCurry Architects*. Last accessed July 27, 2015 from http://www.tigerman-mccurry.com/project/prairie-crossing

1.3 Eco Modern Flats

ArchDaily. "Eco Modern Flats / Modus Studio." *ArchDaily*. June 19, 2012. Last accessed July 27, 2015 from http://www.archdaily.com/245487/eco-modern-flats-modus-studio/

Modus Studio. "Eco Modern Flats." *Modus Studio*. Last accessed July 27, 2015 from http://www.modusstudio.com/project_item/eco-modern-flats/

1.4 Briar Chapel

Eco-Communities. "Briar Chapel Green Community." Last accessed May 6, 2015 from http://www.greenecocommunities.com/North-Carolina/Briar-chapel-green-community.html

Cline Design. "Briar Chapel." Last accessed May 6, 2015 from http://www.clinedesignassoc.com/pages/portfolio/default.aspx?CategoryID=11

2.1 8 House

ArchDaily. "8 House / BIG (Bjarke Ingels Group" 20 Oct 2010." *ArchDaily*. October 20, 2010. Last accessed May 11, 2015 from http://www.archdaily.com/?p=83307

Gonchar, Joann. "8 House – Bjarke Ingels Group." *Architectural Record*. August 2011. Last accessed May 11, 2015 from http://archrecord.construction.com/projects/portfolio/2011/08/8-House.asp

Lomholt. I. "8 House, Copenhagen." *E-Architect*. January 7, 2015. Last accessed May 11, 2015 from http://www.e-architect.co.uk/copenhagen/8-house

2.2 Orenco Station

Costa Pacific Communities. "Orenco Station." *Costa Pacific Communities*. Last accessed July 30, 2015 from http://www.costapacific.com/orenco-station.php

Keller Williams Portland Premiere. "Orenco Station." Orenco Station. Last accessed July 30, 2015 from http://orencostation.com/

Mehaffy, Michael. "Orenco Station." Terrain.org. Last accessed July 30, 2015 from http://www.terrain.org/unsprawl/10/

2.3 Holiday Neighborhood

Barrett Studio Architects. "Holiday Neighborhood." *Barrett Studio Architects*. Last accessed July 30, 2015 from http://www.barrettstudio.com/portfolio/communities-civic/holiday-neighborhood

Boulder Housing Partners. "Holiday Neighborhood Development Project." *Boulder Housing Partners*. Last accessed July 30, 2015 from https://boulderhousing.org/holiday-neighborhood-development-project

2.4 Kuntsevo Plaza

Furuto, Alison. "Kuntsevo Centre / The Jerde Partnership." ArchDaily. November 10, 2012. Last accessed May 14, 2015 from http://www.archdaily.com/?p=287664.

The Jerde Partnership. "Kuntsevo Centre." The Jerde Partnership. 2012. Last accessed May 14, 2015 from http://www.jerde.com/regions/place171.html

LaSalle, Jones Lang. "Kuntsevo Plaza to open in Moscow in August 2014 (RU)." Europe Real Estate. November 12, 2013. Last accessed May 14, 2015 from http://europe-re.com/kuntsevo-plaza-open-moscow-august-2014-ru/43420

3.1 Prospect New Town

Bressi, Todd. "Prospect: Expectations and Enthusiasms [imaginative intensity]." *Places 14(3)*. 2002. Last accessed May 18, 2015 from https://placesjournal.org/assets/legacy/pdfs/prospect-expectations-and-enthusiasms.pdf

Buntin, B. Simmons. "Announcing Unsprawl: Remixing Spaces as Places." *Terrain.org*. 2013. Last accessed May 18, 2015 from http://www.terrain.org/unsprawl/8/

Prospect New Town. "Our Story – Colorado's Premier New Urbanist Community." *Prospect New Town*. 2015. Last accessed May 18, 2015 from http://www.prospectnewtown.com/neighborhood/#our-story

3.2 Masdar City Development

BBC. "Case study: Masdar City in Abu Dhabi." *Urbanisation in MEDCs*. 2014. Last accessed May 2, 2015 from http://www.bbc.co.uk/schools/gcsebitesize/geography/urban_environments/urbanisation_medcs_rev7.shtml

Caine, Tyler. "Inside Masdar City." *ArchDaily*. June 23, 2014. Last accessed May 20, 2015 from http://www.archdaily.com/?p=517456

Foster + Partners. "Masdar Development – Abu Dhabi, United Arab Emirates, 2007." *Foster + Partners*. 2007. Last accessed May 21, 2015 from http://www.fosterandpartners.com/projects/masdar-development/

Kingsley, Patrick. "Masdar: the shifting goalposts of Abu Dhabi's ambitious eco-city." *Wired.co.uk*. December 17, 2013. Last accessed May 21, 2015 from http://www.wired.co.uk/magazine/archive/2013/12/features/reality-hits-masdar

3.3 UniverCity

DIALOG. "Head Master – An intelligent university community takes shape at the top of Burnaby Mountain." *DIALOG*. July 30th, 2015 from http://www.dialogdesign.ca/projects/univercity-at-simon-fraser-university/

UniverCity. "A Natural Attraction." *UniverCity*. Last accessed July 30, 2015 from http://univercity.ca/community/

UniverCity. "A Sustainable Community." *UniverCity*. Last accessed July 30, 2015 from http://univercity.ca/overview/

UniverCity. "Better Living on Burnaby Mountain." *UniverCity*. Last accessed July 30, 2015 from http://univercity.ca/

UniverCity. "Measuring Performance 4 x E = Sustainability." *UniverCity*. Last accessed July 30, 2015 from http://univercity.ca/sustainability/

3.4 Bassin 7 (BSN7)

AARhus. "Byen i bygningen." *AARhus*. 2014. Last accessed May 20, 2015 from http://aarhusiaarhus.dk/

Aarhus Kommune. "BIG og Gel Architects skal gentaenke havnefront." September 13, 2013. Last accessed May 20, 2015. *Aarhus Kommune* from https://www.aarhus.dk/da/omkommunen/nyheder/2013/September/BIG-og-Gehl-Architects-skal-gentaenke-havnefront.aspx

Designboom. "Bjarke Ingels reveals plans for Aarhus island promenade." *Designboom*. September 16, 2014. Last accessed May 20, 2015 from http://www.designboom.com/architecture/big-bjarke-ingels-group-aarhus-island-basin-7-denmark-09-16-2014/

Dezeen Magazine. "BIG plans to 'breathe life' into Aarhus harbour with swimming baths and beach huts." *Dezeen Magazine*. September 25, 2015. Last accessed May 20, 2015 from http://www.dezeen.com/2014/09/25/big-aarhus-harbour-bassin-7-bjarke-ingels/

Rosenfield, Karissa. "New BIG-Designed Neighborhood to Activate Aarhus' Waterfront." *ArchDaily*. September 25, 2013. Architecture AU. "Heller Street Park and Residences." Last updated August 28, 2012 from http://architectureau.com/articles/the-common/

4.1 Heller Street Park and Residences

Architecture AU. "Heller Street Park and Residences." Last updated August 28, 2012. Last accessed May 20, 2015 from http://architectureau.com/articles/the-common/

Inhabit. "Heller Street Park and Residences." Last accessed May 20, 2015 from http://inhabitat.com/heller-street-park-residences-award-winning-sustainable-housing-by-six-degrees-architects-in-melbourne/web-heller-st-04-prodriguez-copy/

Six Degrees Architects. "Heller Street Parks and Residences." Last accessed May 20, 2015 from http://sixdegrees.com.au/projects/residential/heller-street-park-residences

4.2 Arkadien Winnenden

Green Dot Awards. "Ecological City Arkedien Winnenden." Last accessed May 20, 2015 from http://www.greendotawards.com/winners/zoom2.php?eid=3-460-11&uid=2226291&code=Build

Inhabitat. "Arkadien Winnenden is a Family and Earth Friendly Eco Village Near Stuttgart, Germany." Last updated September 18, 2012 from http://inhabitat.com/arkadien-winnenden-is-a-family-earth-friendly-eco-village-near-stuttgart-germany/

Landezine. "Arkadien Winnenden by Atelier Dreiseitl." Last updated April 3, 2013 from http://landezine.com/index.php/2013/04/arkadien-winnenden-by-atelier-dreiseitl/

4.3 Brick Neighborhood

ArchDaily. "Brick Neighbourhood / dekleva gregorič architects." *ArchDaily*. March 10, 2015. Last accessed May 24, 2015 from http://www.archdaily.com/?p=606228

dekleva gregorič architects. "Brick Neighbourhood – 3D erosion to provide identity." dekleva gregorič architects. 2011. Last accessed May 24, 2015 from http://dekleva-gregoric.com/housing-brdo/

4.4 Arbolera de Vida

Design Workshop. "Arbolera de Vida – Albuquerque, New Mexico." *Design Workshop* 2015. Last accessed May 24, 2015 from http://www.designworkshop.com/projects/arbolerade-vida.html

Design Workshop. "Arbolera de Vida Master Development Plan." *Digital Collections*. 1996. Last accessed May 24, 2015 from http://digital.lib.usu.edu/cdm/ref/collection/Design/id/2747

O'Catherine, Aileen. "Does Albuquerque Sawmill Land Trust Work?" *About Travel*. Last accessed May 24, 2015 from http://albuquerque.about.com/od/neighborhoods/i/landtrusts_2.htm

Sawmill Community Land Trust. "Our Projects." *Sawmill Community Land Trust*. 2012. Last accessed May 24, 2015 from http://www.sawmillclt.org/Our-Projects

4.5 Accordia

Feilden Clegg Bradley Studios. "Accordia." *Feilden Clegg Bradley Studios*. 2008. Last accessed May 25, 2015 from http://fcbstudios.com/work/view/accordia?sort=az&direction=asc

Grant Associates. "Accordia." *Grant Associates*. 2008. Last accessed May 25, 2015 from http://www.grant- associates.uk.com/projects/accordia-cambridge/

5.1 Christie Walk

Australian Institute of Landscape Architects. "Case Study #9 – Eco-Village" *Australian Institute of Landscape Architects*. 2006. Last accessed May 25, 2015 from https://www.aila.org.au/

Downton, Paul. "Christie Walk Ecological CoHousing Project." Dr. Paul Downtown, Ecopolis. 2006. Last accessed May 25, 2015 from http://ecopolis.com.au/ecopolis/Christie_Walk.html

5.2 Geuzentuinen

FARO architecten. "Apartment Buildings Geuzenveld." *BNA*. 2003. Last accessed May 25, 2015 from http://www.dutcharchitects.org/projectdetail/2383/

FARO architecten. "Geuzenveld." *Architecture News Plus*. 2003. Last accessed May 25, 2015 from http://www.architecturenewsplus.com/projects/1583

5.3 Parkrand

ArchDaily. "Parkrand / MVRDV." *ArchDaily*. May 1, 2014. Last accessed May 25, 2015 from http://www.archdaily.com/?p=501413

MVRDV. "Parkrand." MVRDV. 2006. Last accessed May 25, 2015 from http://www.mvrdv.nl/projects/parkrand

5.4 Osage Courts

Van Meter Williams Pollack LLP. "Osage Courts." Van Meter Williams Pollack LLP. 2009. Last accessed May 25, 2015 from http://www.vmwp.com/projects/osage-courts-apartments.php

6.1 Artscape Wychwood Barns

Artscape DIY. "Artscape Wychwood Barns." *Artscape DIY.* N.p., n.d. Last accessed May 26, 2015 from http://www.artscapediy.org/Case-Studies/Artscape-Wychwood-Barns.aspx

6.2 Alley 24

Better Bricks. "Alley 24." *BetterBricks*. N.p., n.d. Last accessed May 26, 2015 from http://www.betterbricks.com/design-construction/case-studies/commercial/10054/1

Howard, Sebastian."A New Face for Seattle: A LEED-certified speculative development in Seattle is both environmentally and economically sustainable." *GreenSource*. N.p., n.d. Last accessed May 26, 2015 from http://greensource.construction.com/projects/2009/03_alley24.asp

6.3 Rag Flats

ArchDaily. "Rag Flats / Onion Flats." *ArchDaily*. March 31, 2011. Last accessed May 26, 2015 from http://www.archdaily.com/123621/rag-flats-onion-flats/

Rethink Montgomery."Case Studies – Project: Rag Flats." *Montgomery Planning*., n.d. Last accessed May 26, 2015 from http://www.montgomeryplanning.org/design/rethink_montgomery/casestudies.shtm

6.4 Eden Bio

Etherington, Rose. "Eden Bio by Edouard Francois 2." *Dezeen*. Feb 17, 2009. Last accessed May 26, 2015 from http://www.dezeen.com/2009/02/17/eden-bio-by-edouard-francois-2/

Maison Édouard François."Eden Bio." *Maison Édouard François*. 2009. Last accessed May 26, 2015 from http://www.*edouardfrancois*.com/en/all-projects/housing/details/article/145/eden-bio/#.U3LnovldX7B

6.5 Can Ribas

ArchDaily. "Can Ribas / Jaime J. Ferrer Forés." *ArchDaily.* June 13, 2013. Last accessed May 26, 2015 from http://www.archdaily.com/?p=386089.

Bordas, David. "Ordenació del recinte industrial de Can Ribas." *Publicspace*. 2011. Last accessed May 26, 2015 from http://www.publicspace.org/en/works/g125-ordenacio-del-recinte-industrial-de-can-ribas

Jaime J. Ferrer Forés. "Can Ribas, Palma, Mallorca." *Jaime J. Ferrer Forés*. 2005. Last accessed May 26, 2015 from http://www.ferrerfores.com/?page=41&img01=0

7.1 Via Verde

Grimshaw Architects. "Via Verde." Grimshaw *Architects*. Last accessed June 17, 2015 from http://grimshaw-architects.com/project/via-verde-the-green-way/

Urban Land Institute. "ULI Case Studies: Via Verde – New York City." *Urban Land Institute*. January 16, 2014. Last accessed June 17, 2015 from http://uli.org/case-study/uli-case-studies-via-verde/

"Via Verde / Dattner Architects + Grimshaw Architects." *ArchDaily* March 11, 2014. Last accessed June 17, 2015 from http://www.archdaily.com/468660/via-verde-dattner-architects-grimshaw-architects/

Via Verde. "The Building." *Via Verde – The Green Way*. 2013. Last accessed June 17, 2015 from http://www.viaverdenyc.com/the_building

7.2 Vaudeville Court

"Inventive Council Housing / Levitt Bernstein." *ArchDaily* March 6, 2015. Last accessed June 17, 2015 from http://www.archdaily.com/602610/inventive-council-housing-levitt-bernstein/

Levitt Bernstein. "Vaudeville Court, Islingtom. *Levitt Bernstein*. Last accessed June 17, 2015 from http://www.levittbernstein.co.uk/00,property,796,159,00.htm

7.3 The Beaver Barracks Community Housing

Barry J. Hobin & Associates Architects Incorporated. "Beaver Barracks non profit Housing." *Barry J. Hobin & Associates Architects Incorporated*. Last accessed June 18, 2015 from http://hobinarc.com/projects/portfolio/beaver-barracks/

Sousa, Sandra. "Beaver Barracks and Sustainable Community Development." *Canada Green Building Council*. August 13, 2014. Last accessed June 17, 2015 from http://cagbcottawa.ca/cagbcorc/index.php/read/green-buildings-in-ottawa/50-beaver-barracks-and-sustainable-community-development-site-tour

"The Beaver Barracks Community Housing / Barry J. Hobin & Associates Architects." *ArchDaily*. January 2014. Last accessed June 17,2015 from http://www.archdaily.com/469626/the-beaver-barracks-community-housing-barry-j-hobin-and-associates-architects/

7.4 Landgrab

Etherington Rose "Landgrab City by Joseph Grima, Jeffrey Johnson and José Esparza", January 12, 2010, https://www.dezeen.com/2010/01/12/landgrab-city-by-joseph-grima-jeffrey-johnson-and-jose-esparza/ Accessed September 9, 2017.

7.5 18Broadway

"18BROADWAY : HOME". N.p., n.d. Web. Last accessed May 2, 2014 from http://www.18broadway.com/index.html

Kellner, Jessica. "Kansas City's 18Broadway Urban Rain Garden Showcases Sustainability." *Mother Earth Living*. N.p., Apr. 2011. Last accessed May 2, 2014 from http://www.motherearthliving.com/gardening-tips/kansas-city-18broadway-urban-rain-garden-sustainability.aspx?PageId=1#axzz31Ac2LvOu

8.1 Jubilee Wharf

"Jubilee Wharf." Jubilee *Wharf UK*.Last accessed June 18, 2015 from http://www.jubileewharf.co.uk/about

"Jubilee Wharf." Zedfactory. Last accessed June 18, 2015 from http://www.zedfactory.com/zed/?q=node/116

8.2 Solar Settlement in Freiburg

"Solar Settlement at Schlierberg." *Freiburg-Vauban*. Last accessed June 18, 2015 from http://www.werkstatt-stadt.de/en/projects/22/

8.3 Ironhorse at Central Station

"Ironhorse at Central Station Affordable Housing, All, Apartments, BRIDGE Housing, Green Oakland, California." *David Baker Architects*. Last accessed June 18, 2015 from http://www.dbarchitect.com/project_detail/3/Ironhorse%2520at%2520Central%2520Station.html

Kimura, Donna. "Ironhorse Rolls into Oakland." *Housingfinance*. January 2010. Last accessed June 18, 2015 from http://www.housingfinance.com/development/ironhorse-rolls-into-oakland.aspx

8.4 Österäng

Sesam Arkitektkontor. "Österäng." *Sesam Arkitektkontor*. Last accessed June 18, 2015 from http://www.sesam-ark.se/index.php?page=uppdrag-2

8.5 Solhusen Gårdsten

Low Impact Housing. "Gårdsten Project." *Low Impact Housing*. Last accessed June 18, 2015 from http://sealevel.ca/lowimpact/housing/action.lasso?-Response=search05.lasso&ID=1380

"Rejuvenating a Social Housing District." *SymbioCity*. Last accessed June 18, 2015 from http://www.symbiocity.org/Templates/Pages/Page.aspx?id=133&epslanguage=en

World Habitat Awards. "Solar Housing Renovation, Gårdsten Project." 2005. Last accessed June 18, 2015 from http://www.worldhabitatawards.org/winners-and-finalists/project-details.cfm?lang=00&theProjectID=293

9.1 Donnybrook Quarter

Markasaurus. "Housing Showdown: Donnybrook Quarter and Robin Hood Gardens." *Markasaurus*. February 2010. Last accessed June 22, 2015 from http://markasaurus.com/2010/02/10/housing-showdown-donnybrook-quarter-and-robin-hood-gardens/

Peter Barber Architects. "Donnybrook Quarter." *Peter Barber Architects*. Last accessed June 22, 2015 from http://www.peterbarberarchitects.com/01_Donny.html

9.2 Living Places Suburban Revival

2012 Melbourne Design Awards. "Living Places Suburban Revival." *2012 Melbourne Design Awards*. Last accessed June 22, 2015 from http://melbournedesignawards.com.au/mda2012/entry_details.asp?ID=10963&Category_ID=4763

Bent Architecture. "Living Places Suburban Revival." *BENT ARCHITECTURE*. Last accessed June 22, 2015 from http://www.bentarchitecture.com.au/projects/completed-works/living-places-suburban-revival.html

9.3 Greenwich Millennium Village Phase 02

Housing Prototypes. "Millennium Village Phase 2." *Housing Prototypes*. Last accessed June 22, 2015 from http://housingprototypes.org/project?File_No=GB004

Proctor and Matthews. "Greenwich Millennium Village/Phase 2 – London SE10." *Proctor and Matthews*. Last accessed June 22, 2015 from http://www.proctorandmatthews.com/sites/default/files/pdfs/projects/92-greenwich-millennium-village.pdf

9.4 Cottages on Greene

U.S. Department of Housing and Urban Development. "East Greenwich, Rhode Island: Cottages on Greene's Innovative Approach to Infill." *U.S Department of Housing and Urban Development*. Last accessed August 5, 2015 from http://www.huduser.org/portal/casestudies/study_07022012_1.html

Union Studio. "Cottages on Greene." *Union Studio*. Last accessed August 5, 2015 from http://unionstudioarch.com/projects/cottages-on-greene/

10.1 Columbia Station Micro-community

Dwell Development. "Columbia Station (Phase One)." *Dwell Development*. Last accessed June 23, 2015 from http://dwelldevelopment.net/columbia-station/

Hickman, Matt. "Evergreen Homes: Columbia Station." *MNN*. February 14, 2012. Last accessed June 23, 2015 from http://www.mnn.com/your-home/remodeling-design/blogs/evergreen-homes-columbia-station

10.2 SOL Austin (V1.0)

SOL Austin. "Modern, Sustainable House Plans." *SOL Austin*. Last accessed June 23, 2015 from http://www.solaustin.com/Green_Building_2-1_m1090.php

SOL Austin. "Net-Zero Energy Formula: Conservation, Efficiency, and Integrated Systems Approach." *SOL Austin*. Last accessed June 23, 2015 from http://www.solaustin.com/net-zero_energy_formula.php

10.3 Sweetwater Spectrum Community

ArchDaily. "Sweetwater Spectrum Community / LMS Architects" *ArchDaily*. Last accessed June 24, 2015 from http://www.archdaily.com/446972/sweetwater-spectrum-community-lms-architects/

Sweetwater Spectrum. "Environmental Stewardship." *Sweetwater Spectrum – Life with Purpose*. Last accessed June 24, 2015 from http://www.sweetwaterspectrum.org/environmental_stewardship.aspx

10.4 Paisano Green Community

ArchDaily. "Paisano Green Community / WORKSHOP8." *ArchDaily*. September 20, 2012. Last accessed June 24, 2015 from http://www.archdaily.com/271384/paisano-green-community-workshop8/

WORKSHOP8. "MEDIA KIT – Paisano Green Community." *WORKSHOP8*. August 7, 2013.Last accessed June 24, 2015 from http://www.hacep.org/public/upload/files/resources/Paisano_Green_Community_Media_Kit.pdf

WORKSHOP8. "Paisano Green Community Senior Housing." *WORKSHOP8*. Last accessed June 24, 2015 from http://workshop8.us/portfolio/paisano-green-community-senior-housing/

10.5 zHome

David Vandervort Architects. "zHome." *David Vandervort Architects*. Last accessed June 24, 2015 from http://vandervort.com/multiprojects/zhome/

David Vandervort Architects. "zHome certified by International Living Future Institute." *David Vandervort Architects*. June 11, 2013.Last accessed June 24, 2015 from http://vandervort.com/news/zhome-certified-by-international-living-future-institute/Irina Vinnitskaya. "UPDATE: zHome / David Vandervort Architects." *ArchDaily*, March, 30 2012. Last accessed June 24, 2015 from http://www.archdaily.com/220740/update-zhome-david-vandervort-architects/

照片版权信息

Photography Credits

1.1 Civano Moule & Polyzoides, Architects & Urbanists

1.2 Prairie Crossing Courtesy of the architect

1.3 Eco Modern Flats Modus Studio (page 22); Timothy Hursley (pages 23, 24); Adaptive Creative (page 25)

1.4 Briar Chapel Newland Communities

2.1 8 House Courtesy of the architect

2.2 Orenco Station Courtesy of the architect

2.3 Holiday Neighborhood Barrett Studio Architects

2.4 Kuntsevo Plaza Courtesy of the architect

3.1 Prospect New Town DPZ CoDESIGN

3.2 Masdar City Development Foster + Partners (page 60); Nigel Young/Foster + Partners (pages 61–5)

3.3 UniverCity SFU Community Trust

3.4 Bassin 7 (BSN7) Courtesy of the architect

4.1 Heller Street Park and Residences P. Rodriguez

4.2 Arkadien Winnenden Rolf Messerschmidt (pages 83 top, 84); Studio Dreiseitl (page 83 bottom)

4.3 Brick Neighborhood Miran Kambic

4.4 Arbolera de Vida Dale A. Horchner

4.5 Accordia Tim Crocker

5.1 Christie Walk Dr. Paul Downton, Architect

5.2 Geuzentuinen Tomei (page 106); Luuk Kramer (page 107); IC4U – Hans Peter Föllmi (pages 108 bottom left, 109); Kees Hummel (page 108 right)

5.3 Parkrand Rob t' Hart

5.4 Osage Courts Van Meter Williams Pollack

6.1 Artscape Wychwood Barns Tom Arban (pages 123, 124 left, top right, middle right, 125); DTAH (page 124 bottom right)

6.2 Alley 24 Timothy Soar (pages 127 top, 128 top right); Lara Swimmer (page 127 bottom); Gummi Brynjarsson (page 128 left); Benjamin Benschneider (pages 128 bottom right, 129)

6.3 Rag Flats Raimund Koch (pages 131, 132 top right, 133); Onion Flats (pages 132 left and bottom right)

6.4 Eden Bio Nicolas Castet (pages 135, 137 bottom left), Alex Dahl (page 136); David Boureau (page 137 top)

6.5 Can Ribas ESTOP (page 138); José Hevia (pages 139–41)

7.1 Via Verde David Sundberg

7.2 Vaudeville Court Tim Crocker

7.3 The Beaver Barracks Community Housing Doublespace Photography

7.4 Landgrab Jeffrey Johnson (page 159); Dezeen (pages 160, 170)

7.5 18Broadway John Iiams (pages 162, 164 top, 165); DST (page 164 bottom right)

8.1 Jubilee Wharf Julian Calvery

8.2 Solar Settlement in Freiburg Rolf Disch Solar Architecture

8.3 Ironhorse at Central Station Brian Rose (page 179); DavidBaker Architects (page 180); Steve Proehl (page 181)

8.4 Österäng Sesam Arkitektkontor AB

8.5 Solhusen Gårdsten Christer Nordström

9.1 Donnybrook Quarter Courtesy of the architect

9.2 Living Places Suburban Revival Trevor Main

9.3 Greenwich Millennium Village Phase 02 Courtesy of the architect

9.4 Cottages on Greene Nat Rea/Nat Rea Photography

10.1 Columbia Station Micro-community Tucker English

10.2 SOL Austin (V1.0) Chris Krager (page 221); Chandler Prude (pages 222, 223)

10.3 Sweetwater Spectrum Community Tim Griffith

10.4 Paisano Green Community Jesse Ramirez

10.5 zHome Aaron Ostrowsky

建筑师信息
List of Architects

Barrett Studio Architects
1944 20th St, Boulder, CO 80302, United States
www.barrettstudio.com
bsa@barrettstudio.com
+1 303 449 1141

Bent Architecture
Level 1, 14 Wilson Ave, Brunswick VIC 3056, Australia
www.bentarchitecture.com.au
info@bentarchitecture.com.au
+61 03 9388 9033

BIG (Bjarke Ingels Group)
Kløverbladsgade 56, 2500 Valby, Copenhagen, Denmark
www.big.dk
big@big.dk
+45 7221 7227

Boulder Housing Partners
4800 N. Broadway, Boulder, CO 80304, United States
www.boulderhousing.org
bhpinfo@boulderhousing.org
+1 720 564 4610

BRIDGE Housing
San Francisco office
600 California St, Suite 900 San Francisco, CA 94108, United States
www.bridgehousing.com
+1 415.989.1111

Cline Design
125 N. Harrington St, Raleigh, NC 27603, United States
www.clinedesignassoc.com
office@clinedesignassoc.com
+1 919 833 6413

Costa Pacific Communities
11416 SW Barber St, Wilsonville, OR 97070, United States
www.costapacific.com
info@costapacific.com
+1 503 646 8888

Dattner Architects
1385 Broadway, NYC, NY 10018, United States
www.dattner.com
info@dattner.com
+1 212 247 2660

David Baker Architects
461 Second St, Loft c127, San Francisco, CA 94107, United States
www.dbarchitect.com
info@dbarchitect.com
+1 415 896 6700

David Vandervort Architects
2000 Fairview Ave E, Suite 103, Seattle, WA 98102, United States
www.vandervort.com
online form at http://vandervort.com
+1 206 784 1614

Design Workshop
Asheville Office
60 Biltmore Ave, Asheville, NC 28801, United States
www.designworkshop.com
dwi@designworkshop.com
+1 828 280 9637

dekleva gregorič architects
Dalmatinova Ulica 11, SI-1000 Ljubljana, Slovenia
www.dekleva-gregoric.com
arh@dekleva-gregoric.com
+386 1 430 52 70

DIALOG
Vancouver office
406-611 Alexander St, Vancouver, BC V6A 1E1, Canada
www.dialogdesign.ca
general@dialogdesign.ca
+1 604 255 1169

DPZ CoDESIGN
Miami, Florida Office
1023 SW25th Ave, Miami, FL 33135, United States
www.dpz.com
online form at http://www.dpz.com/
+1 305 644 1023

DTAH
50 Park Rd, Toronto, ON M4W 2NS, Canada
www.dtah.com
admin@dtah.com
+1 416 968 9479

Dwell Development
4501 Rainier Ave S., Seattle, WA 98118, United States
www.dwelldevelopment.net
dwell@dwelldevelopment.net
+1 206 683 7595

Ecopolis (Dr. Paul Downton)
9 / 681-683 Nepean Highway, Carrum VIC 3197, Australia
www.pauldownton.org
paul@ecopolis.com.au
+ 61 8 8410 9218

FARO architecten
Estate Olmenhorst, Lisserweg 487d, 2165 as Lisserbroek,
The Netherlands
www.faro.nl
info@faro.nl
+31 (0) 252 414 777

Feilden Clegg Bradley Studios
London Office
20 Tottenham St, London W1T 4RF, United Kingdom
www.fcbstudios.com
London@fcbstudios.com
+44 (0)20 7323 5737

Foster + Partners
London Office
Riverside, 22 Hester Rd, London SW11 4AN, United Kingdom
www.fosterandpartners.com
+44 (0)20 7738 0455

Grant Associates
22 Milk St, Bath BA1 1UT, United Kingdom
www.grant-associates.uk.com
info@grant-associates.uk.com
+44 (0) 1225 332664

Grimshaw Architects
637 West 27th St, New York, NY 10001, United States
www.grimshaw-architects.global
+1 646 293 3600

Hobin Architecture Incorporated
63 Pamilla Street, Ottawa, ON K1S 3K7, Canada
www.hobinarc.om
mail@hobinarc.com
+1 613 238 7200

HOK
New York Office
1065 Avenue of the Americas, 6th floor, NYC, NY 10018, United States
www.hok.com
newyork@hok.com
+1 212 741 1200

Jaime J. Ferrer Forés
07012 Palma de Mallorca, Illes Balears, Spain
www.ferrerfores.com
jaimeferrer@coaib.es

Jeffrey Johnson
45 Main St Suite 824 Brooklyn, NYC, NY 11202, United States
www.slab-a.net
slab@slab-a.net
+1 718 666 3330

The Jerde Partnership
Los Angeles Headquarters
913 Ocean Front Walk, Venice, CA 90291, United States
www.jerde.com
matt.heller@jerde.com
+1 310 399 1987

Jose Esparza
josesparza@gmail.com

Joseph Grima
joseph@grima.net

KRDB
916 Springdale Rd, Austin, TX 78702, United States
www.krdb.comjoseph@grima.net
info@krdb.com
+1 512 374 0946

Leddy Maytum Stacy Architects
677 Harrison St, San Francisco, CA 94107, United States
www.lmsarch.com
info@lmsarch.com
+1 415 495 1700

Levitt Bernstein
1 Kingsland Passage, London E8 2BB, United Kingdom
www.levittbernstein.co.uk
hello@levittbernstein.co.uk
+44 (0)20 7275 7676

Margaret McCurry FAIA, Tigerman-McCurry Architects
444 North Wells St, Suite 206, Chicago, IL 60654, United States
www.tigerman-mccurry.com
tma@tigerman-mccurry.com
+1 312 644 5880

Maison Édouard François
7 Passage Thiéré, Paris, France
www.edouardfrancois.com
maison@edouardfrancois.com
+33 (0)1 45 67 88 87

Modus Studio
15 N. Church Ave #102, Fayetteville, AR 72701, United States
www.modusstudio.com
contact@modusstudio.com
+1 479 455 5577

Moule & Polyzoides, Architects & Urbanists
180 East California Boulevard Pasadena, CA 91105, United States
www.mparchitects.com
info@mparchitects.com
+1 626 844 2400

MVRDV
Rotterdam Office
Achterklooster 7, 3011 RA Rotterdam, The Netherlands
www.mvrdv.nl
office@mvrdv.com
+31 (0)10 477 2860

NBBJ
Seattle Office
233 Yale Ave N, Seattle, WA 98109, United States
www.nbbj.com
seattle_office@nbbj.com
+1 206 223 5555

Nordström Kelly Architects AB
Asstigen 14, 436 45 Askim, Sweden
www.nordstromkelly.se
hej@nordstromkelly.se
+46 031 282 864

Onion Flats
111 West Norris St, Philadelphia, PA 19122, United States
www.onionflats.com
info@onionflats.com
+1 215 426 6466

PacTrust
Corporate Office
15350 SW Sequoia Parkway, Suite 300
Portland, OR 97224, United States
www.pactrust.com
+1 503 624 6300

Peter Barber Architects
173 King's Cross Rd, London WC1X 9BZ, United Kingdom
www.peterbarberarchitects.com
info@peterbarberarchitects.com
+44 (0)20 7833 4499

Proctor and Matthews Architects
7 Blue Lion Pl, 237 Long Ln, London SE1 4PU, United Kingdom
www.proctorandmatthews.com
info@proctorandmatthews.com
+44 (0)20 7378 6695

Rolf Disch Solar Architecture
Sonnenschiff, Merzhauser Str 177, Freiburg 79100, Germany
www.rolfdisch.de
info@rolfdisch.de
+49 0761 459 44 0

Sesam Arkitektkontor AB
Cardellsgatan 8, 291 31 Kristianstad, Sweden
www.sesam-ark.se
+46 044 18 48 80

Six Degrees Architects
110 Argyle St, Fitzroy VIC 3065, Australia
www.sixdegrees.com.au
info@sixdegrees.com.au
+61 3 9635 6000

Union Studio
140 Union St, Providence, RI 02903-1714, United States
www.unionstudioarch.com
info@unionstudioarch.com
+1 401 272 4724

Van Meter Williams Pollack LLP
1738 Wynkoop St, Suite 203, Denver, CO 80202, United States
www.vmwp.com
www.vmwp.com/contact
+1 303 298 1480

Wayne Moody
(1939–2005)

WORKSHOP8
1720 15th St, Boulder, CO 80302, United States
www.workshop8.us
+1 303 442 3700

ZED Factory
21 Sandmartin Way, Wallington SM6 7DF, United Kingdom
www.zedfactory.com
info@zedfactory.com
+44 (0)20 8404 1380

索引

Index

A

adaptable housing 190–1

Australia
- Christie Walk 102–5
- Heller Street Park and Residences 78–81
- Living Places Suburban Revival 198–201

C

Canada
- Artscape Wychwood Barns 122–5
- The Beaver Barracks Community Housing 154–7
- UniverCity 66–9

carbon-reduction strategies 6, 210–13

China
- Landgrab 158–61

community gardens 76–7, 92, 142–4, 145, 156, 209

compact neighborhoods 98–101

connectivity 30–3, *see also* mobility networks

D

Denmark
- 8 House 34–9
- Bassin 7 (BSN7) 70–3

design
- compact neighborhoods 98–101
- environmental considerations 10–13
- historical development 9, 118–21
- innovative dwelling concepts 190–3
- micro-communities 98–101
- mixed-use development 40–3, 44–7, 52–5, 60–5, 70–3, 146–9, 170–7
- mobility networks 8, 30-2, 34, 76
- passive design strategies 10–11, 222–3
- urban environments 8

district heating 166–9, *see also* geothermal strategies

E

edible landscapes 142–5, *see also* permaculture

energy production 166–9

environmental considerations 10–13, *see also* green open spaces, historical development, sustainability

F

farming, urban 142–5

France
- Eden Bio 134–7

G

geothermal strategies 154, 167, 168, *see also* district heating

Germany
- Arkadien Winnenden 82–5
- Solar Settlement in Freiburg 174–7

green open spaces 74–7, *see also* community gardens, edible landscapes

H

heating systems 166–9, *see also* passive solar strategies

heritage preservation 118–21

historical development 118–21

housing, innovative concepts 190–3, *see also individual themes*

I

infill housing 9, 118–21

infrastructure planning 30–4, *see also individual themes*

innovative dwelling concepts 190–3

L

LEED certification 22, 26, 114, 124, 128, 224, 228

low-carbon residences 210–13

M

micro-communities 98–101

mixed-use
- development 52–5, 60–5, 146–9, 170–3
- neighborhood 40–3, 44–7, 70–3, 174–7

mobility networks 8, 30–2, 54, 76

N

Netherlands, The
- Geuzentuinen 106–109
- Parkrand 110–13

neighborhoods, mixed-use 40–3, 44–7, 70–3, 174–7

new urbanism 40–3, 56–9, 124

P

parks, *see* green open spaces

passive design 10–11, 222–3

passive solar strategies 10–11

photovoltaic panels 16, 190, 200, 237
  solar panels 76, 104, 181

permaculture 14–17

planning, *see* urban planning

photovoltaic panels 16, 190, 200, 237

public transit 31–3, 44, 54, 66, 99–100

R

renovated housing 22, *see also* adaptable housing

roofs, green 85, 133, 146, 153, 156, 162

Russia
  Kuntsevo Plaza 48–51

S

Slovenia
  Brick Neighborhood 86–9

solar panels 76, 104, 181

solar settlements 174–7

Spain
  Can Ribas 138–41

sustainability, *see also* new urbanism
  carbon reduction strategies 6, 210–13
  energy production 166–9
  low-carbon residences 210–13
  passive design 222–3
  passive solar strategies 10–11
  solar settlements 174–7
  permaculture 14–17

Sweden
  Österäng 182–5
  Solhusen Gårdsten 186–9

T

transit, public 31–3, 44, 54, 66, 99–100

U

United Arab Emirates
  Masdar City Development 60–5

United Kingdom
  Accordia 94–7
  Donnybrook Quarter 194–7
  Greenwich Millennium Village Phase 02 202–5
  Jubilee Wharf 170–3
  Vaudeville Court 150–3

United Nations Conference on the Human Environment 7

United States
  18Broadway 162–5
  Alley 24 126–9
  Arbolera de Vida 90–3
  Briar Chapel 26–9
  Civano 14–17
  Columbia Station Micro-community 214–19
  Cottages on Greene 206–9
  Eco Modern Flats 22–5
  Holiday Neighborhood 44–7
  Ironhorse at Central Station 178–81
  Orenco Station 40–3
  Osage Courts 114–17
  Paisano Green Community 228–33
  Prairie Crossing 18–21
  Prospect New Town 56–9
  Rag Flats 130–3
  SOL Austin (V1.0) 220–3
  Sweetwater Spectrum Community 224–7
  Via Verde 146–9
  zHome 234–7

urban density 8, 98–101

urban farming 142–5

urban planning 8
  community gardens 76–7, 92, 142–4, 145 156, 209
  compact neighborhoods 98–101
  district heating 166–9
  edible landscapes 142–5
  infill housing 9, 118–21
  innovative dwelling concepts 190–3
  mobility networks 8, 30–2, 34, 76
  public transit 31–3, 44, 54, 66, 99–100
  urban density 98–101
  urban farming 142–5
  urban revival 141, 149
  waste management 142–3
  water management 68, 142–3, 162, 224–7

W

waste management 142–3

water management 142–3
  18Broadway 162
  Sweetwater Spectrum Community 224–7
  UniverCity 68